WATER MAMAS

A Novel of Climate, Spirituality, and Indigenous Human Rights

Mona Shomali

Book Cover Artwork by Mona Shomali

For Ian

ACKNOWLEDGMENTS

Thank you to Dayna Bennett, April McMurtry, Stephanie Moon, David Godfrey, Emily Henderson, and Mae Chan.

FOREWORD

Afa didn't know whether to be scared or allured by her beauty. Her scales shimmered in the moonlight, and her silky wet hair glinted with beads of moisture. Her eyes shone with something Afa couldn't make out from here, but it intrigued her.

"Come here," the woman whispered. "I want to speak to you."

Afa stared across to where the woman clung to the banks of the river, her hands gripping the sandy edged basin. Was she a woman though? Afa could see her face, her hands... but her torso she could barely make out in the dark night. Afa squinted, trying to get a better look. But a heavy mist crept in and enshrouded the woman, obscuring Afa's view.

"Why? Why should I come over there? What do you want?" Afa called out.

"There is something I need to tell you, something that you need to know," the woman–or whatever she was–called back.

Afa waved her hand through the air, trying to clear the mist so she could once again see the woman.

“Do I know you?” Afa asked. She looked around, suddenly wondering where she was. Above her, a dark tree, laden with branches and leaves of darkest green. Under her feet, sand. The jungle. She was back in the Amazon.

“You must know me, Afa.”

The voice now so close, a warm breath could be felt on Afa’s cheek and she gasped. She stumbled backwards, falling into the soft leaf litter.

ONE

Afa shifted uncomfortably in her linen pants, slapping repeatedly at her legs as insects bit through the thin cloth. Was linen really the best choice for the wet rainforest? She wondered, glancing around the room at the tribal leaders. Nothing could have prepared her for the humidity. Her hair stuck to the back of her neck and was curling, sticky around her temples. At least the nervous sweat mixed with sunblock and insect repellant wouldn't be obvious since her shirt was completely damp anyway. Her presentation to the Macushi tribe was to begin any minute, and she couldn't quell the nerves despite the repeated box breathing. Inhale for four, hold for four, exhale for eight. She was an outsider, and she assumed she was not welcome. Despite years of therapy in New York City, the critical voice of her father intruded on her thoughts and beat down her confidence in times like these.

"Elders, may I have your attention," Thomas, an elderly tribal leader from the Macushi tribe, addressed the group of indigenous leaders from Guyana that had gathered to hear Afa speak. The tribal elders were seated in a circle, hot and crowded, in a thatched roof building made of dried plant materials and wood. Afa noticed that the leader that addressed the group was wearing a button up short-sleeved cotton shirt and some khaki shorts with sandals. His cell phone was in the pocket of his shirt. Everything about him looked to be of Western influence – except for the semi-permanent colorful markings on his face. Many tribal members in the room wore the same markings which came from dyes sourced from local roots found on their land. Others wore ceremonial headdresses or other regalia reserved for showing rank, for important meetings, something distinctive to set them apart as the tribal elders. But most were dressed similarly to Thomas, the elder introducing her, and were it not for the facial markings, could have passed for locals on the streets of New York City. Everyone was on their cell phones, but several looked up to greet her when she entered, and some stood to shake her hand. Still others barely acknowledged her. Everyone was speaking in English since Guyana had been colonized by the British, and everyone's first names, with the exception of some Macushi names, sounded like British names.

Thomas waited for the group to give him their attention before continuing. "Please join me in welcoming our guest, Afa, from the United Nations. Afa is here to talk to us about the United Nation's rain program. She is seeking our consent to begin the program to bring back rain to the Amazon."

Afa wanted so much to be her professional and confident self, but even though she'd just gone to the bathroom, nervous energy made her feel like she was going to go in her pants. Get focused! She said to herself, reaching for a glass of water and putting it to her lips, hoping the cool liquid would ease the flush rising in her cheeks. Afa flicked her long straight black hair out of her face, then adjusted her sweaty bra. Rubbing her chin on her apple shaped face, she squinted her almond eyes and looked thoughtfully at the elders in the room. She hoped she looked approachable and non-threatening.

When Thomas had mentioned the rain program, Afa noted several grumbles had emitted from the crowd. But this was why she was here – this resistance. She hoped to put their minds at ease and answer all their questions. One or two tribal members smiled at her; several met her gaze with a distrustful stare or stoic look; still others appeared to scowl. Afa felt her mouth go dry again. Resisting the urge to slap at the insects, she stood and held her arms open, hoping her voice rang out as "self-assured" and "trustworthy" to this crowd.

"Esteemed Macushi tribal Elders, thank you so much for hosting me today. I am so pleased to be here. I know there is a lot to go over, and we have a long day ahead of us. But it is my goal to give you all the information you need to make an informed decision and give consent to this rain program."

Afa stressed the word "informed," leading with her intent. Trying to gauge the energy of the room, she sensed an interesting mix of hostility and curiosity, so she pressed forward. "As you all know, what used to be a large continuous South American rainforest is now only a series of small patches of rainforest left intact and inhabited by Indigenous Peoples like you, where you have cared for your forests. Anthropologists have long proven that where Indigenous Peoples inhabit and manage the Amazon, the forests are most preserved, and the ecosystem is at its healthiest. You are excellent managers of the rainforest. But your remaining patches are like a little oasis surrounded by deforestation, dead trees, and drought. Each remaining patch of Amazon jungle is unique because each still has rain, but the United Nations wants to bring rain to the whole Amazon ecosystem, which will not only make your patches bigger, but will also benefit the whole planet as the Amazon once was instrumental in regulating global climate. As you may be aware, the Amazon rainforest once recycled much of the annual rainfall back into the global atmosphere and regulated global climate. How can we generate rain and bring it back? It is brought in through rockets and the use of rain generating chemicals. But what is most exciting is that we now have the power to directly impact and potentially reverse climate change in our lifetime..."

"Yeah, by shooting rockets full of poison into the air!" A stern-faced man heckled her from the back of the room. Several voices rose up in assent, and the members of the tribes began murmuring to one another. Afa froze and took a minute to regain her composure – she'd expected resistance, but she couldn't say she'd expected to be heckled. While the tribal members shifted in their seats and the grumbling continued, she noted several of the tribal elders were nodding at her, as if to give her the go ahead. Taking a deep breath, she held up a hand and waited for the chatter to cease, then pressed on.

"I understand you have concerns about the chemicals, and I am here to explain what chemicals are used in the rockets. And I know I'm an outsider and you have a history of outsiders attempting to manage your land. So first off, I'd like to tell you a little about myself and my qualifications – and why I truly believe this rain program is in your best interest." Afa waited to make sure the members were following and went on once she felt she had their attention. "However, I don't plan to do all the talking today. I also want to hear from you; what are your concerns? What do you hope to gain from this program? And what do you need from me, from the UN, to feel satisfied in allowing this program to proceed?"

The room had fallen silent. She could hear the batting of the straw fan on the ceiling, beating the humid air through the room. She had never felt so much moisture in her lungs. Or between her breasts, for that matter.

"I am Afa, and while I may live in New York City now, my family is from a small island in the south pacific that doesn't exist anymore. Our island disappeared as a result of sea level rise brought about by climate change. I lost my home, and as a young girl, I was powerless to stop it. So, as you can imagine, I am sensitive to the fact that your patches here in the Amazon are getting smaller. And I know you have lost so many plants, trees, and animals that were a part of the bigger jungle. I can't emphasize enough the impact of deforestation on the global community. The Amazon rain cycle has important impacts on global climate..."

One of the tribal elders put his hand up and began to speak without waiting for Afa to finish.

"Ms. Afa, we agree that deforestation has affected our culture and way of life. We have had so little power to stop deforestation. As the forest disappears, so do we," he said with a soft voice that sounded like agreement.

"Yes, I understand. I was powerless as a little girl to stop my island from disappearing. But all of us in this room, we are no longer powerless. We have the tools and technology to create rain! I understand you may have questions. Allow me to go over the details of the program, and I promise you there will be time to address each and every one of your concerns."

The heckler from earlier started to speak, but an elder cut him off abruptly, "Let her speak." Then, turning to Afa, he said, "Go ahead."

Afa breathed a sigh of relief and wiped at the sweat that was dripping into her eyes. More than anything, Afa needed the communities to feel comfortable, to understand her great respect for them. Without all the information, she feared they would reject the program outright because of fear. She hoped she could gain their trust and convey the benefits before the heckler shut her down again. As much as Afa believed in the program and its far-reaching benefits for the entire world, she was beginning to feel she had to work a lot harder than she'd expected to get the buy-in she needed from these tribal communities. Looking at the elder with a nod of gratitude, Afa continued her presentation.

"I work in the Climate Modification Department at the United Nations where we initiate large scale multi-country rain programs all over the world. Right now, countries and states are all launching their own weather programs to create rain, but we oversee programs that cross borders, where international agreements must be signed and ratified. Currently, the Chinese rain program is the biggest program in the world, and China has been able to use rockets to create rain for a region twice the size of the American state of Alaska. To address food shortages and the agricultural crisis, China's rain programs are now able to schedule rain and recreate seasons for farmers. The Beijing Weather Modification Office has been able to create rain for 60% of China. However, American rain rocket intelligence has been catching up to China, with the state of California now operating the second largest program after China. The Californian government has created rain programs aimed at the drought-ridden hardwood forests that have become a tinderbox of yellow pine needles, dry branches, and trunks already hollowed out by fire. The rain schedules in California mimic the rain cycles from 50 years ago, the baseline year – which is amazing! Seasons have become reliable again!" Afa's voice rose and her eyes gleamed with passion, her hands animating her words.

The California rain program had been an amazing success, ending years of drought driven wildfires that had practically decimated the state.

"The Amazon program, if the agreement is signed by all the state governments and Indigenous Peoples that inhabit the patches, would be the largest multi-country program. The Amazon was once the world's largest rainforest, where the majority of the world's species lived, so the outcome of this program would affect the whole planet." Afa grinned at the tribal community before her, uncertain as to whether she had sold the benefits of the agreement to them as best she could. She personally believed there was just no other way to look at the rain program, other than good. Good for the people, good for the planet. They just had to see it that way.

An elder who hadn't spoken before stood up, his chest heaving with effort. Afa looked at him expectantly, hopefully.

"Yes? Sir?" she asked.

"Why do you want our consent? It's not like you ever asked us before! This is all just talk. We have rain on our patches, we do not need you!" His chest puffed up and down.

Afa's smile slipped from her face. Why had she assumed this would be easier?

"Of course, that's an important question. As you are aware, Indigenous Peoples living in the patches must give their consent. International Human Rights instruments such as the Inter-American Court of Human Rights have ruled that Indigenous Peoples must give "FPIC," or Free, Prior, and Informed Consent of any large scale projects that are initiated on their land. Your consent is crucial. The United Nations seeks FPIC of Indigenous Peoples to start the rain program. And I understand that you still have rain on your patches, some of the last rainforests preserved on earth. But the patches are getting smaller every day, and without bringing back rain to the entire Amazon, your patches will disappear. That is why I am presenting to you today with some urgency, to secure your participation in the program and agreement."

Here, Afa paused once more, trying once again to read the room. She was relieved the heckler from earlier had stayed silent while she spoke. The next part would be the hard part. Readying herself, she asked, "Any other questions so far?" While Afa waited for the elders to confer, she fidgeted nervously with the woven mat covering the lectern where she stood. She took a moment to observe the space, noting that familiar blossom of doubt that was rooted in her gut, the omnipresent feeling of not being good enough. Doubt in herself, her work, her presentation – it sat there, threatening to disrupt the competent appearance she worked so hard to present. She knew that there was a symphony of qualities that spoke to her being good at her job, but the insecurity in her ability to do well was sometimes stronger. When that low opinion of herself threatened to overtake her, all she could do was push back at the memory of her dad's relentless criticism, which she had somehow learned to direct at herself.

An elder who had been preoccupied with his phone suddenly shoved the device into his chest pocket, shifted around in his seat and shouted out, "This is all well and good, but how does this all work?" He swiveled back to face Afa. "How do you actually make the rain?" He was wearing a feathered headdress and had black markings on his face but otherwise wore a plain white button down shirt, the cell phone peeking out above his front pocket.

"Of course, and thank you for asking that question. I am not a scientist myself, but I'll do my best to explain. The technology is called cloud seeding. Chemicals are shot out of machines – called artificial rain enhancement rockets – over large areas of land. The most common chemicals used in cloud seeding are silver iodide, potassium iodide, dry ice, and liquid propane. The chemicals are then dispersed, and the particles make clouds which create rain..."

A younger woman with jet black hair softly interrupted, "These chemicals, do they cause harm? We have heard that silver iodide is toxic. Can you please talk more about that before you continue?"

"Absolutely. While some studies have shown that silver iodide can cause temporary injury in humans and the ecosystem, that's only with long term exposure. But you will not have long term exposure. So far, in seeding programs in California, the amounts of silver iodide are not even large enough to register. In other words, the amount of silver iodide used is harmless. There are many detailed studies and extensive research that shows negligible environmental impacts."

The same woman spoke up again, kindly pressing, "But Ms. Afa, we have read about Australia, how people have been resisting the rain program because of the silver iodide. If it is so safe, can you explain what is happening in Australia?" She leaned in eagerly, as if giving Afa a real chance to explain.

"The rain program in Australia was rejected by the government because of environmental concerns. They cited harmful effects on a protected species. However, the claims of negative environmental impact were disputed by scientists at the International Weather Modification Association. But Australia was never able to restart their program because the public resistance was so great. I believe the people in Australia who fought the rain program were misinformed. They rejected a program that would have really helped them – they are in a 50 year drought!" Afa paused for emphasis. "Places where rain programs have been successful, people are happy and relieved. The rain programs have had a good effect on people's mental health because the expectancy of rain is good for people. The benefits so far outweigh the risks – the programs are improving the world, restoring ecosystems, and the chemicals are not toxic in the amounts that are used."

The room was quiet, and Afa wasn't entirely sure if they had taken in her response. She noted that some of the younger leaders had seemed to be tuning her out, their faces angled down at the screens in their hands. After a few seconds, Thomas, who appeared to be one of the most senior and respected leaders in the room, raised his hand to speak. Afa nodded in his direction.

"Ms. Afa, can we participate in the scheduling of the rain? Will we be notified of when the rain would begin and for how long would it continue?"

"That is a great question Thomas, and to be honest, I'm not sure. The rain is scheduled based on studying global circulation models – or GCMs – from the baseline years where there was regular rainfall, so recommendations and predictions are made based on these parametric models."

"Will the rockets that shoot the rain seeds into the air scare our animals?" asked one of the older men. He spoke slowly and stared intently at Afa, waiting for a response.

"Another good question. I don't know how loud the rockets will be, but that is a question that I can ask when I get back to the United Nations and I can respond to you later. Is that ok?"

"Yes, Ms. Afa, that is ok with me."

The questions continued for another hour and Afa did her best to answer them, often deferring to the scientists back home, saying she would have to get back to them when she returned to New York. She felt frustrated at her lack of scientific knowledge; why hadn't she anticipated more of these concerns and come ready with answers?

Although they did not have to give her an answer today, once the presentation ended, she asked Thomas "How do you think it went?"

Thomas looked back at the other indigenous leaders behind him and with a steady gaze he told Afa, "We will need to go back to our tribes and discuss everything, of course, before we can give our final consent. I can't say yet how they'll feel about signing on to the rain agreement."

Afa nodded slowly and felt hopeful, despite his guarded response. This was probably the best outcome she could have expected for today. So much of this rain agreement was riding on this community trusting her; that Thomas hadn't dismissed her outright when she asked felt like a little vote of confidence.

"Ms. Afa, we have arranged for you to meet with Noel, the Director of the International Science Station managed by Macushi on Macushi land. We want you to also meet Vishnu, the biologist we work with, before you go back to Georgetown and on to New York. They will tell you more about our patch. We will speak to you again once we have met with our tribes. Thank you for coming to see us." Thomas smiled warmly at Afa and grasped her hand in his. Afa noted that his smile reached his eyes, which had tiny creases of life all around them.

"Thank you, Thomas." Afa returned Thomas's smile. "I look forward to seeing you again, too." She turned her gaze out the doorway to the thick, darkly matted jungle, speckled with light from openings in the trees above, pierced by a single snake-like muddy red road only passable by way of jeep or truck. She would drive along this jungle road to the capital city, Georgetown.

She remembered looking out the window of the tiny twin engine plane headed for Georgetown, watching the small South American country of Guyana come into view. She had flown from New York to Venezuela the night before where she'd met her connecting flight to the capital city, which lay along the North Atlantic coast. Afa felt like she was walking into a furnace as she emerged from the airport. The high-pitched sound of insects was deafening, and she had wondered if they were cicadas, or some other insect unknown to her. And here in the Amazon it was so cool in contrast to the bustling heat and crowd of the concrete and trash lined city of Georgetown, a coastal city that was mostly paved, but where there were occasional red dirt roads. The ocean water that met the city walls was muddy and full of sediment, not suitable for swimming. Georgetown had a legacy of Dutch and English colonizers, and the architecture reflected that, unlike the sister capital cities of South America that were colonized by the Spanish. The wooden structures had uniquely designed Dutch windows, jutting out of the building, very specific to Guyana. There were few high rises, and few trees. It seemed to be a place where deals were made, where international bureaucrats came to do business and left.

Afa sat waiting, sweat gathering between her legs, until finally a yellow van pulled up and woman in a tight floral dress with a long, black braid descended and began to walk towards her. Her huge hoop earrings glinted against her caramel complexion as she waved and called out.

"Afa? Are you Afa?" Her voice carried a melodic Caribbean accent.

"I am! Are you Noel?" Afa responded, standing and hoping her pants weren't sticking to the sweat between her legs.

"Yes! How was your flight and your drive out to the Amazon?"

Noel reached out to take Afa's bag from her shoulder and pointed towards the van.

"It was fine. I slept most of the time."

"That's lucky. I can't sleep on flights. I'm a nervous flyer," Noel said, laughing as she opened the door of the van. "Let me tell you about my relationship to the Macushi till Vishnu arrives. He will ride the eight hours back to Georgetown with you. It's a long trip – we won't be arriving till morning. It will be hard to sleep with all the bumps, but you can try."

To this day there is only one road that connects Georgetown to the interior Amazon communities. It is unpaved, muddy and frequently impassable during the rainy season.

Afa nodded grimly. She was familiar, and her one trip from Georgetown into the jungle the day before wasn't far enough in the past to have faded from memory.

"Have you used the restroom? You know there is only one stop at an outpost about four hours into the drive. Otherwise, it's straight on through. The road is very bumpy, as your bum knows." She winked. "Sometimes the car gets stuck in the mud in the patch, but most of the drive is dry dirt in the post rainforest savannahs."

"I just used the restroom, thank you, so I should be okay. And despite the bumps, I'm looking forward to the drive," said Afa, patting her rear. She was struck by how energetic Noel was, and she had liked her gregarious and upbeat nature right away.

"My dear woman, you're looking forward to being in this car for eight hours? I could never say the same thing!" said Noel with laughter. "Why don't we sit in the van and become acquainted with one another," said Noel. They got into the van and tried to make themselves comfortable amongst the boxes.

"How long have you been Director of the International Science Station on the patch?" asked Afa.

"About five years. I go back and forth between the patch and Georgetown. Our capital city once had a beautiful coast, but that was a long time ago. You will drive by the sea wall when you arrive in Georgetown in the morning and see how sea level rise has flooded the homes on the coast. They're all empty now," said Noel with a sad shake of her head. "Terrible."

"Devastating." Afa shook her head in agreement. "I'm originally from the South Pacific Island of Tuvalu that was submerged. I am very familiar with sea level rise."

"Oh, really? Have you become American? Did you change your nationality?"

"Yes, we all did. Our passports became meaningless once the island was submerged."

"Afa, I am sorry for your loss," Nicole said sincerely, looking at Afa. "I cannot imagine what it is like to lose your country. I love Guyana and I couldn't live anywhere else. I'm a real mixed culture myself. My mother is Indo Guyanese, and my father is Afro Guyanese. Both Indians and Africans were brought to this country as servants and slaves, of course."

"Such an interesting background. I did some research on Guyana and I hadn't realized how mixed the people are until I read about it," said Afa. "Did you get all your education here in Guyana?"

"No, I got my undergraduate degree in Canada but returned to Guyana for my PhD. Then I got my job at the science station."

"Do you like your work? It seems like a fascinating job. I bet you meet people from all over."

"I love it! And yes, I meet all the scientists and students that come from all over the world. We even have international study abroad students coming through and that's probably my favorite part. I love working with them." Noel smiled with her eyes as she spoke.

"I think that's what makes your patch so unique. When I was reading about the work you're doing here, it was the partnership with the indigenous people that stood out to me. Was it tricky to gain their trust, to structure the partnership with the Indigenous Peoples?" asked Afa.

"Oh, I am happy to tell you about that!" Noel's enthusiasm for the subject was apparent by the broad grin that spread across her face. "This is what makes the Macushi patch so wonderful. We are the only patch in the Amazon with ecotourism and a science station that is managed by Amerindians – what we here in Guyana call the Indigenous Peoples. The Macushi both maintain the land restoration and manage the industry on the patch. We are a very special patch." Noel said this with evident pride.

"How many indigenous tribes are on the land?"

"Just the Macushi. The other tribes have left to live on another patch closer to Brazil, but some have stayed behind and joined the Macushi. The tribal leader of the Macushi is Marcia, and she is a force. She rarely leaves the Science Station and Eco Tourist facilities. That is why she was not here today to meet you. Marcia has lived on the patch her whole life. She grew up watching her father use his ecological knowledge to reintroduce the Arapaima fish back into the river after the fish was thought to be extinct, and as a teenager she worked in ecotourism. She worked her way up to be the director of the ecotourism program until she was elected as the leader of the Macushi people."

"She sounds like an amazing woman. I'm looking forward to meeting her."

"Yes, you will like her. Everyone does," said Noel with a look of absolute confidence. "You will meet her next time you come." Then with a side-long glance she said, "You are coming back, right?"

"I am coming back!" said Afa, trying to match Noel's enthusiasm.

The driver, an older man with deep set lines all over his kindly face, had been sitting patiently while Afa and Noel spoke. He suddenly began waving his hand ecstatically. Afa looked to see who he was so excited about and noticed a good-looking young man approaching the van.

The man was tall and slim, wearing a baseball cap with sunglasses perched on top and was very casually dressed in an olive-green t-shirt, camel colored cargo shorts and brown sandals. With his deep brown skin, he looked Indo Guyanese, part of the large population of Indians that were brought over as slaves by the British as Noel had explained. Younger than her by maybe ten years, he had a relaxed, mischievous smile. She saw him quickly assess her from head to toe before his gaze finally rested on her eyes, and he reached through the window of the van to shake her hand.

"Hi. I'm Vishnu. You must be Afa, with the UN. You're here to discuss bringing the rain program to the Amazon, right?"

"Yes, that's me." She shook his hand, slightly unnerved by his intense gaze, not to mention his good looks.

Noel got out of the van to greet him, excitedly yelling, "Vishnu is here!" Noel had emailed Afa that she would be riding back to Georgetown with a Guyanese biologist who was based out of the compound at the science station. She'd noted that it was rare to have a Guyanese biologist on the station since most of the scientists were foreign. What Noel hadn't mentioned was that the biologist would be quite so striking and charming.

"Well, let's get going," Vishnu said as he climbed in next to Afa in the beat-up van. The air conditioning was turned on as the car engine began its grumble, a stark contrast to the oven-like humidity of the Amazon. Despite being under the canopy shade of the layered forest of many colors and shades of leafy green, it was still damp warm air.

"Noel tells me you're working closely with the Indigenous Peoples on this rain project." Vishnu looked over at Afa as the van began its bumpy journey along the jungle roads.

Afa nodded. "That's right. I'm meeting with the Macushi to see if they will agree to sign a United Nations agreement to bring rain to the entire Amazon region. Hopefully this will expand the patches, and the forests will grow and expand again."

"I work with the indigenous too," said Vishnu while nodding his head. "My PhD is focused on indigenous ecosystem management, preserving the way that Indigenous Peoples in Guyana perceive their environment."

"Really…that's an interesting focus. How does that work, exactly, with the Macushi?" Afa asked, intrigued by the premise of his focus. She also found the sound of Vishnu's voice intriguing; it seemed wild yet tamed at the same time.

"Mainly, I'm documenting their world views and how that informs how they manage their land, what plants they use as medicine. I'm also documenting the restoration programs they have engineered. For example, they have brought species that were on the brink of extinction. They were able to do it with their deep knowledge of the river ecosystem."

"Incredible!" Afa was genuinely impressed and found it impossible to hide it. "So you work alongside the scientists too, or just the Macushi?"

"Just the Macushi. I spend a lot of time with Marcia, the director, and the botanist who is also a shaman, as well as the elders. These villages have democratically elected toshaos who are village leaders. So I meet with them and try to capture all their stories. And, I meet with the young people who have formed a "rangers" program for young ecosystem protectors. Most of the young people want to move to the big cities like Boa Vista in Brazil and Georgetown, but they eventually come back when they want to start families."

Afa's body suddenly flew up off her seat, then landed hard as the van rolled over a particularly massive rut. She looked out the window and saw the wet red dirt, the color of rust, with the giant holes they could not avoid on the uneven road. She looked at Vishnu, who did not seem rattled in the least by the bumps. "So, this is your PhD research? Macushi spirituality and spiritual world views? Is this, like, folklore...myths, legends, that type of thing?" She leaned in, hoping the long car ride would give her a chance to learn more. And a chance to forget about the bumps.

"I think so. Westerners tend to see Indigenous Peoples like the Macushi as conservationists, managing their ecosystem sustainably, but in reality, while they do in fact have traditional environmental knowledge and a close relationship to the forest species and animals, most of the management comes from a spiritual perspective that they pass down as social and cultural memory. The "folklore" you mention. The unintended consequence of that spiritual view is what western scientists would consider and prove to be good and sound ecosystem management. Their cultural and spiritual views mean that they manage their natural resources sustainably. Cultural management is the modality of management, not science. The majority of the spiritual stories have to do with traditional beliefs and taboos that form their identity. They consider themselves managers of the forest, or forest stewards who are guided by spirits who do not let them test the limits of their forest resources. Following the will of the spirits means that they have used their resources within the limits of regeneration, they don't over-harvest or degrade their environments. So conservation is not an active process, it is an unconscious result, or byproduct, of everyday spiritual perspective."

"What do you mean by spiritual perspective? Can you tell me what it is?" asked Afa.

"Macushi creation stories do not center around a single god, nor do they believe in or worship a single god. They see their world as inhabited by spirits. They believe that there are guardian spirits, or masters in the forest, masters that govern certain species, and tributes must be made to the masters so that the community remains healthy. There are also evil spirits that must be avoided. For example, the Macushi know the areas of the forest that are considered sacred or inhabited by "evil spirits" and they do not enter these locations to hunt. All the plants, trees, animals, the water, the sky – all have spirit masters. The Macushi have relationships with the spirits, and there are taboos to avoid angering the bad spirits – or the oma – which bring illness and death, so they always had to be on guard. The taboos may ban access to resources during certain time periods. Taboos may be imposed daily, weekly or seasonally. For example, the communities may protect a body of water from fishing for a certain time because of a taboo."

"Do they fear the spirits?"

"It's not so much that they fear them, it's that they listen to them. For instance, it is believed that various spirits live in the trees, and when entering the forest for an activity such as hunting, extracting something from the trees or farming, the tribe need to prepare themselves."

"So how would they prepare?" Afa asked, riveted.

"They talk to the trees, and only the shaman can say which trees can be cut, and how to cut the tree to avoid damaging the tree spirit. The Macushi stories say that if the cutting of a tree was not done right, the guardian spirits of the trees could destroy and bring bad consequences to the people. In this way, key species of trees are protected. Because of their beliefs about the consequences of angering the spirits, in the past tribes only took what they needed, guided by the shaman. The taboos and fears of repercussion are internalized and affect the way people view their environment and the resources they use."

"Okay, that makes an interesting sort of sense, doesn't it?" Afa nodded.

Vishnu nodded and smiled in agreement. "It does. But as the Macushi have been exposed to western ideas, the youth rely less on the stories. The elders, naturally, believe that it is their following of the spiritual beliefs that has kept their patch intact all these years while the rest of the Amazon around them has burned, dried up or has been logged. The elders fear that the youth are not learning the stories and rituals used to pay tribute to the spirits. And for now, the elders are teaching me." Vishnu paused as if considering his next words. "There is a stereotype that indigenous people live in harmony, like the ecologically noble savages who live on the surface of the land in a culturally primitive way, which is praised by western environmentalists. This puts an unfair cultural burden on young people who want to leave the forest. It is like non-Indigenous Peoples view them as…" he searched for a word. "Less indigenous, I suppose, for wanting to lead a western life." Vishnu paused and seemed suddenly reluctant to go on. "By the way, please stop me if I am boring you. It's not hyperbole to say I could talk about this all day. It's been the focus of my PhD for years." He laughed.

"No, please, continue. I'm truly fascinated. Tell me more about the shaman. I think it might be relevant to my work. The shaman sounds like he or she has a lot of power."

Vishnu looked pleased. "Happily, then." He bowed his head. "The botanist for the Macushi, who is also the shaman, is known by the Macushi as the Piaimen who gives his blessing when it is time to extract resources, based on his or her relationship with the spirits. The Piaimen knows all the stories – they are the spiritual leader. He or she is the intermediary between the spirits and the tribe. Like a "medicine man" – or woman – who can speak with the spirits and help cure the sick, provide protection, that sort of thing. This person is like the spiritual channel that can address any imbalance in the environment through communication with spirits. With their spiritual knowledge, the community is directed in how to behave for balance to return. The Piaiman shapes the rules, norms and taboos followed by the tribe to guard against the actions of spirits and protect their resources. The shaman can remove the oma, or the bad spell which is interpreted as a consequence for not taking care of resources."

Afa, who'd been listening thoughtfully, interjected, "I can't help but consider this and how it relates to my own task here, with the Macushi. Do you think I should seek out the shaman and speak to them about the rain program?"

"No, you should speak to everyone. The Macushi now, because of the ecotourism on their patch, all make decisions with the ecotourists in mind. Things have become more complicated as outsiders stay on their patch."

"So what do you think my biggest challenge will be in discussing the rain program?"

"I can imagine your work with the Macushi would be difficult because of their belief in Kanaima, a type of shape-shifting evil spirit that is personified as an "outsider" that brings death and misfortune to the tribal members. This outsider can be seen as a dark shaman. You are clearly an outsider and so you must work against that belief."

"Wow! I...I didn't know about, what was it? Kananima?" Afa was astounded.

"Kanaima," Vishnu corrected. "And yes, but you are talking about bringing rain which would be good for their ecosystem, so I am not sure how they would respond to the UN. Perhaps you will have Marcia's ear, which will be useful."

"I'm looking forward to meeting her."

"Yes, she is very influential and involved in every part of the operations of the science station, as well as respected among the tribe, of course. If this is your first trip to the patch, may I ask how it went when you visited with the Macushi?"

"Well, I honestly don't know. They communicated with me that they are skeptical of the program, but that they would have their final answer on whether they want to join the agreement on my next trip here. When I come back, I'll meet again with community members and hear what they have to say. So far, I have learned that all the elders believe that the Macushi patch is the most successful patch in the Amazon. Now, after talking to you, I believe that perhaps because of their adherence to their spiritual beliefs and their culture, they may be skeptical of the silver iodide chemicals that are launched out of the rockets for cloud seeding not so much for their own health risks, but because it might interfere with their spirits."

"True, but more accurately, they may fear that the chemicals will specifically affect their water spirits, which are now called 'Water Mamas.' Kind of like the western notion of a mermaid, they govern the water. The water spirits are known to steal and save souls, to attract and repel men and women, to entice people into destruction. And just like with mermaids, there are notions of sexual attraction between humans and Water Mamas. Water Mamas can also capture people and turn them into Water Mamas. One of the elders told me that deaths amongst the Macushi are described in terms of a person finding a spouse under the water, or a union with a Water Mama. The Water Mamas are sometimes thought to be master owners of fish or rivers, or sometimes both. So I can very much imagine that the Macushi will be worried about their Water Mamas."

"Wow. Water Mamas." Afa pictured a long haired, scaly, mermaid-like creature emerging from the banks of an Amazonian river, with swirls of mist encircling her as she beckoned. "I'm truly stunned, Vishnu. Of course I needed to know this before meeting with the Macushi. I am so glad I am talking to you now. Most of my preparation and briefing on the Macushi from the UN had to do with the ecotourism and the Science Station and how it's been so successful. There was almost nothing in my notes about the Macushi spiritual worldview. But of course! It's everything." Afa sat there, shaking her head, realizing what a big part of the puzzle she had missed.

"You know," she said to Vishnu. "Prince Harry came to visit in the 2010s and he was a supporter of the Macushi. They have lots of celebrities flying in, landing on their helicopter pad."

"That's the kind of thing that was in my briefing documents." She shook her head again, frowning at the ineptitude of her own organization.

"Ah yes, they do have famous visitors. There is a helicopter pad right on the patch for the uber wealthy tourists. But that doesn't get to their heads. They are mostly concerned with keeping the forest alive and wet, and growing cassava as that has special symbolism in their culture."

"Cassava, that is an ingredient in the Macushi dish, pepperpot? I read about that."

"Have you?" Vishnu smiled broadly. "Yes, the cassava is reduced to a thick, black, sweet syrup that is used to make traditional foods. The root is a staple crop of the Macushi and has been for over 3,000 years. In the old days, only women prepared the cassava root, extracting the poison and transforming it into the edible form, but now men and women prepare the cassava, and on the ecotourism tour, ecotourists are shown the process of extracting the poison."

"Are you also learning about all the plants in the forests too?" asked Afa.

"Yes, I am trying to document all the medicinal plants on the patch."

"Truly fascinating, Vishnu."

"I'll tell you more about that later. You must be tired of hearing me speak by now!" laughed Vishnu.

"Not at all. I am so grateful to have met you, and to have learned so much," Afa said. Afa couldn't imagine ever tiring of learning about these fascinating people, and she was thinking it had been a bit of a stroke of luck to be thrown in with Vishnu. She might otherwise have missed some majorly vital information about the Macushi had she not. And talking with Vishnu for the entire trip had made the eight hours pass quickly.

They finally arrived in the city where the majority of architecture was European in the main streets, and makeshift exposed cinder blocks held together by clumpy cement in the side streets. It was all against the backdrop of an oppressive heat. The van came to a stop and Vishnu got out first, holding out his hand to help Afa. She took his hand and stepped out onto the concrete. The air here was much drier than it had been in the Amazon, and Afa stretched and immediately asked for the bathroom.

TWO

Back in New York, Afa ordered a cappuccino at a Swedish cafe called Fika. She took her delicate white porcelain cup to a small table near the window, admiring the pale wood furniture and sparse decor. She was meeting with her new boss for the first time and though she typically felt nervous and anticipatory in these situations, she found she was feeling what she'd call "neutral" – not really nervous or excited, just sort of regular. All she knew about the man was what she had received in an email introduction. His name was Kwame and he was from Oakland where he was a director in the Climate Modification program in the state of California. In his initial email to her, he had suggested that they have their first one on one meeting more informally, at a cafe near the building. Maybe the informality of his email had put Afa at ease, but she was expecting a low-key meeting. Through the glass window, she saw a tall black man enter the cafe, and he looked like man in the thumbnail from the email.

So this is my new boss, thought Afa. He was wearing a white hoodie sweater with a blazer on top, and colorful sneakers that she saw young people wearing on the subway. He smiled in her direction, walked over and put his hand out.

"Hi, you must be Afa. I'm Kwame. It's so nice to meet you. Thanks for meeting with me."

"Hi, Kwame. Yeah, of course. I thought it was nice to get out of the office for a bit."

"So you like this cafe? It's a bit posh!" Kwame gestured at the tables.

"Really?" Afa made a quizzical face and looked around. She didn't see what was particularly "posh" about a Swedish cafe.

"Yeah, but I prefer the simpler things. Gas station coffee works fine for me," said Kwame.

Afa was slightly miffed at his insinuation that she was "posh," whatever that meant anyway. This was New York City, and this was not posh by the standards of the city. Were they supposed to meet at a gas station? Annoyed, and feeling like they had gotten off on the wrong foot, she attempted to redirect her agitation and changed the topic. Not generally a fan of small talk, in some situations, like this one, it was obligatory.

"So you came here from Oakland? Are you from Oakland originally?" asked Afa.

"Well my parents are from Ghana originally, but I grew up in England briefly, and then we came to Oakland. Not a typical childhood, really. My dad was a psychiatric nurse in England, and we lived on the grounds of an experimental mental hospital where nurses worked with patients tending to large vegetable gardens. When the hospital closed, we came to the U.S. So yeah, Ghana, England and then Oakland."

"Wow. That does sound like a unique childhood."

"Yeah, I lived in Europe until my mid-twenties. After I graduated college, I started my career doing climate research in England. We were running simulations and predicting climate, working on models. It was great. Then I moved back to Oakland a few years later, and my parents moved with me."

"So what exactly were you doing in California before you came to the UN?"

"Mostly summarizing global circulation models – or GCM's – in order to make recommendations on scheduling the rain programs." Afa was familiar with global circulation models, which were estimations and predictions of climate put out by different meteorologist offices globally. "Basically, I broke down parametric models that predicted local climate, analyzed them, and created rain schedules. We simulated spring, fall and winter rains from our baseline year," said Kwame.

"That sounds incredibly technical," said Afa. She wondered how such a technician would fare amongst the bureaucratic UN where any decision was a huge process with multiple stakeholders who had their own agendas. You need serious people skills to be successful at the UN, she thought. She decided to say something positive, despite her misgivings. "It sounds like you have lots of experience with climate modification," said Afa.

"Yes, I knew from the start that I wanted to work with climate modification. It is what gives my life purpose." Kwame looked thoughtful. "Do you know Paulo Coelho?"

"The Brazilian author? He wrote The Alchemist."

"Yeah. He has a quote. I always butcher it, so this is paraphrasing, but he said, When you become an instrument of the universe, to help the soul of the world, you realize who you are. I love that quote. That's kinda how I feel about the work we are doing at the UN."

"Ooooh, I like that. I'm going to try to remember that." Some of Afa's earlier irritation gave way to a more open attitude towards him. She observed his appearance, very serious expression against smooth skin and high cheekbones.

"So tell me about you, your background. Like, your name – where is it from?" asked Kwame.

Afa felt his eyes on her, as though he was critically examining her appearance, and Afa immediately flushed. She'd worn an oversized puffy coat to fight off the New York chill, and her hair was wild. Suddenly self-conscious that she did not look professional and hadn't taken the time to fix herself up, she attempted to smooth the wild hairs that flew in every direction. "Afa means hurricane in the South Pacific island where I am from."

"Oh, very cool. Which island?" Appearing keenly interested, Kwame leaned in.

"Tuvalu. It's submerged now from sea level rise. We went underwater shortly after the Marshall Islands and the Maldives. But I'm sure you know this – you work at the UN."

"Yes, I am aware. I am sorry, Afa," said Kwame in a gentle tone.

"Well, anyways." Afa was slightly thrown off by the way he'd segued from critical stare to empathetic care. "Your name is also interesting. Any special meaning?" She tried to reset the tone by returning the curious gesture.

"It means born on Sunday. Not as exciting as a hurricane!" Kwame laughed, his gaze intent on Afa. Looking down at the cappuccino cup in her hand, he gestured and said, "I'd offer to buy you a cup of coffee, but I see you already have one."

"Yeah, thanks. I got here a few minutes early."

"So…how did you end up at the UN?"

"I applied through a special program for climate refugees. I qualified, of course, since my island home went under water years ago when I was a child and we moved to New York."

"Oh, got it. That must have been traumatic for you and your family." Afa sensed his empathy, and once again, she found herself unsure of what to make of Kwame.

"It was traumatic, and it basically informed my career. When I first was hired by the United Nations, I was working with sea level rise. I had lived through it and all I wanted to do was help the next island that was disappearing. Then, I started bouncing around different departments and I ended up working on the rain agreement."

"How long have you been working on rain programs?" he asked.

"Just five years."

"Are you enjoying your work in the Amazon? How's it going so far? I read in your notes that you made the presentation in Guyana."

"I did, and they were very hesitant. The Macushi elders are discussing it now with the community on the patch. I am scheduled to go back to Guyana to hear their final decision, after I go to Ecuador to meet with the tribes there."

"Hesitant, huh? Well, that's why we are here. To soothe their concerns. This is an amazing program," said Kwame. Afa wondered if he was aware of the mythologies of Indigenous Peoples and the important role they played in their culture. Is this the right time to tell him they fear that the chemicals will hurt their spirits? she wondered. But she decided to wait. They were just introducing themselves and she didn't want to dive in too deep with him yet. She wanted to get a read on him first. He presented himself like a scientist, a technician, and she didn't know if he had contemplated the social and cultural reasons that a tribe wouldn't sign onto the agreement. She didn't know if he had ever worked directly with the indigenous communities. She didn't know if he just rejected all cultural beliefs in the name of science, or if he saw value in the role of culture when it came to managing the ecosystem. After all, she herself had only just learned of the importance of the spiritual practices and the impact on the Macushi worldview. This needed to be approached delicately, she intuited.

"What's it like for you to experience that amount of natural rain on your visits to the Amazon patches?" asked Kwame with true enthusiasm.

"Oh!" Afa clutched her heart. "It rained every few hours! And the rivers were full – not as full as they were in the baseline years 30 years ago, but still a sight to see. I wish everyone could experience it!"

"Oh, I know," Kwame agreed. "I am in awe of places that still have natural rain. I wonder what places like California were like when they had regular and natural rain. About 10 years ago, I was researching how rain gradually stopped in California because of dieback, but nothing is as severe as the dieback in the Amazon."

Afa nodded vigorously as he was speaking. She knew a lot about Amazon dieback; where the rainforests began to self-destruct once 40% of forest had been removed. Anyone working in climate modification would be passionate about the issue, understanding the urgent need to reverse its effects.

"Seeing dieback in person was really hard. I saw it all through my drive from Georgetown to the patch. I had been reading how the patch is shrinking, how the Indigenous Peoples have lost so much land due to dieback, and seeing it was harder than reading about it." Her hands were flat, splayed across the top of the table as she presented her experience to Kwame. "But when I'm on the patch, I feel optimistic that we can bring the wet rainforest back." Her hope was a force that felt tangible to her.

"Sounds like you are the perfect person to be doing your job; we need this kind of enthusiasm and passion behind us. I'm assuming most of your presentation is procedural. I'm sure that with you on the job, the tribes will sign onto the agreement."

"Well, though I appreciate the vote of confidence, Kwame, I'm not so sure. The Macushi were apprehensive about the chemicals used in cloud seeding, so it might be a bit more difficult than I initially thought."

"Sometimes we have to work with misinformation on the ground. It reminds me of American people who refused to get the vaccine during the 2019 pandemic. If they had an education, if they understood the science, they would get the COVID-19 shot. The uneducated refused and often died in greater numbers. I'm sure you can address the concerns they have about the chemicals. Go in armed with information and educate them."

Afa didn't think that was a fair comparison, and this frustrated her deeply. His comment seemed flippant, irresponsible. There were other reasons people might choose not to get the vaccine, like the trauma and history of healthcare practices in low-income communities in the U.S., or for various other health reasons – it wasn't always a lack of education. She hesitated briefly, resisting the urge to confront her new boss. It wasn't in Afa's nature to hold back – she was feisty, full of life, quick to laugh but also quick to correct someone if they were out of line. If someone pissed her off, they would know about it. People didn't spend a lot of time guessing where they stood with Afa, but the stakes felt high here – too many lives depended on how she handled her job. She wondered if she could gently disagree with his vaccine comparison. Then she squared her shoulders and threw caution to the wind; she knew full well what she brought to the table, and this guy needed to be set straight. She was the one having conversations with the Macushi, not him. The lines between her eyes deepened as she thought of what to say next.

"Respectfully, I'm going to have to disagree. Indigenous Peoples are quite educated about what affects them. They have their own systems of knowledge. As for my job, I take their input really seriously. And when you consider the history of Indigenous Peoples and how they have been left out of decisions affecting their land – it's criminal!" Afa leaned into her words, animated as she drove her point home. Kwame physically backed up from her just a bit, as if expecting her to lunge forward.

"Yes, of course, I didn't mean to suggest that they were wholly uneducated." He held his hands up in surrender, backtracking. "Excuse the bad analogy. I appreciate your passion. That is a quality that I look for in the people who work for me."

Afa felt her shoulders relax, assuaged by his quick apology, and she realized she had been leaning forward across the table at him. She settled back into her chair and re-crossed her legs. "Well. Yes. I get pretty fired up about representing communities I work with. And I'm so looking forward to working in the Amazon. I'm leaving for Ecuador to meet with multiple tribes who share a patch in a few weeks. I'm sure you saw that in the notes."

"Yes, I did. Like I said earlier, it sounds like you are the right person to work with the tribes, gaining approval, and looking for buy-in. The whole world benefits if rain returns to the rainforests, and I am sure the indigenous people realize that. I'm confident you will be successful in gaining their consent."

"Thanks. I do appreciate your support, and I'll do my best, of course. And, I wanted to mention, I'm thinking of hiring a graduate intern from NYU to help me with the science, focusing on the chemicals used in the cloud seeding process and the impact of those chemicals on the ecosystem," said Afa.

"Wait, why are you doing that when we have our own scientists we work with through the UN? How can we be sure this won't promote conspiracy theories about the chemicals in the rockets?"

"We can be sure because I'm considering an intern whose work is focused on attempting to prove that cloud seedling chemicals are not harmful to the ecosystem. I just want a fresh mind on this, someone that is exposed to all the newest science that they are teaching in school. I felt if there was any weakness in my presentation to the communities, it was that I had to ask them to wait for information. It would be better if I had that knowledge prepared for me ahead of time, as much as is possible."

"Okay..." Kwame considered her point. "Okay. I guess it couldn't hurt." He glanced at his watch. "Listen, let's keep chatting while we walk back to the office, shall we?" They stood, and Afa gathered her bag, rushing to keep up with him as he moved towards the door, not wasting any time or seeming to wait for her, continuing the conversation. "So. Moving forward, I expect you to work very hard. I hope I can mentor you as well as your previous manager did, whom I hear you got along with really well. I heard they switched to another department," commented Kwame, and Afa noted he didn't hold the door for her, but just expected her to keep up.

Afa felt in two minds. She knew this type of aloof and technical scientist, and though she wanted to like him, she also wondered how close to the ground he was, in the trenches. She hoped he was not the kind of manager that didn't understand the challenges of the people who report to him. He could be brusque, but overall, her read of him was that he was a good person who was very focused. He believed that his work mattered and loved his job, and she could work with that.

***

In New York, when fresh snow has just fallen, everything is soft powder and white, only broken up with deep footprints. The snow is then shoveled to the sides of the sidewalk and soon the powder white turns grey and slushy at the corners and edges from the traffic. The pure white snow will eventually have yellow spots where people let their dogs urinate. Afa had observed such urination and wondered if the dogs were dehydrated based on the yellow-ness of the color. She often checked her own urine color to see if she was drinking enough water, but the New York city tap water tasted so bad so she didn't drink as much as she should. She had just read a report in the New York Times that there were traces of antidepressants and anti-anxiety medications found in New York city tap water.

Afa was eagerly waiting for her friend Aliah in the snow at the entrance to the pool in Chelsea. They swam indoors together twice a week no matter what the weather. The pool felt familiar to Afa, the smell of chlorine, the sound of splashing water and the scratchy white gym towels. She had swum a lot as a kid on the island, and although this was lap swimming and not the ocean, she was still submerged in water. It felt like a respite from the rest of her day. However, there was always some drama at the pool between the lanes, which were labeled slow, medium and fast. Afa once saw the lifeguard blow the whistle and break up an all-out fight between swimmers over who should be in the fast lane and who should be in the slow lane. Besides these few and far between skirmishes between swimmers, Afa found swimming to be serene and meditative. She could pull the water towards her, moving her body forward on an axle from side to side, in a rhythm so steady that she could lose thoughts and just be in her physical body. It was a rare place that she felt self-assured and confident wholly in herself.

Aliah finally came bouncing around to the front door of the gym in her snow boots and olive green puffy down jacket, her curly black hair escaping from under her wool hat.

"I'm here! Sorry I'm late – some knucklehead on the subway was preventing the doors from closing – it was so crowded on the train," Aliah said as she laughed. Afa always allowed for Aliah to be late and almost expected it. Aliah was fun and spontaneous and untamed, so Afa could graciously accept that her lateness was just one aspect of her personality that went with those other qualities. While talking and embracing as if they had not seen each other for years, they walked into the locker room, took off their clothes and changed into their bathing suits. They pulled out their colorful swim caps and held their goggles in their hands as they walked out onto cold white tiles surrounding the slightly heated pool. It was their tradition to sit down on the side of the pool and kvetch about the day and catch up with what was going on in their lives.

"Is that man with the open sore getting into the pool? You see him? Oh my god, I hope not. Oh god, I think he is! He should not be allowed to get in this pool!" said Aliah. Afa laughed at Aliah's passionate disapproval.

Afa had met Aliah five years ago through a Brazilian Samba dance class at the Alvin Ailey dance center in the Hell's kitchen neighborhood. They were both always at the back, exhausted, struggling to keep up with the girls 10 or 20 years younger than them. Aliah was a dedicated schoolteacher at a public school in the West Village, where the parents were all super wealthy, successful and educated. Aliah and Afa eventually stopped going to the dance class when they realized that they both swam and that they could go to the pool together for exercise, rather than suffer through Samba. They quickly bonded over being active and enjoying the endorphins that came after a strenuous physical workout.

"How was your trip? We haven't really gotten into it since you have been back. What's going on with your rain program? Are the people in the Amazon interested?" asked Aliah.

"Well…to be honest…they were not so hot on the rain program," said Afa, frown lines appearing on her otherwise youthful face. "In a nutshell, they fear the chemicals that are used to make rain."

"What? Why? Are the chemicals really that bad? It'll bring rain to the Amazon. Isn't voting against rain voting against their own interests?"

"Well, it's a bit more complicated. Their culture has very specific folklore around spirits that govern the environment. Like, they have these water spirits, and their belief in them plays a part in a decision that has local impact. It's a dilemma for the global nature of the environmental agenda. This is the classic tension between global environmental science elite and local mythology."

"Hmm." Aliah looked thoughtful. "Well folklore is very local in nature. We did a unit on folklore in my class last year. I heard a lot of stories from around the world that I hadn't heard before, mostly having to do with relationships between people and nature in the places where people lived. I remember one myth that my Bolivian student presented. I think it was called Pachamama? I hope I'm saying it right. It's this indigenous Bolivian belief that all humans, animals, plants, trees are inside the womb of the Pachamama, a goddess-like entity. Everything in her womb is equal. Humans don't have more value than animals and plants, for example. I'm pretty sure this student got help from his parents, because he included in his report that the former president of Bolivia, Evo Morales, assigned animals and trees their own legal rights, the same legal rights as humans." Aliah grinned. She often recollected stories about her students, her pride in them purely visible.

"Huh. That's interesting. My first thought is how different that is from Judeo-Christian writings in the bible that humans have dominion over everything on earth, all the animals, plants, etc. In Christianity, we are on top. This is so different from indigenous cultures who defer to a spirit body that they are a part of, that all life is a part of."

"Yes. So different. I wonder if something like water spirits exists in other folklore. The common thread throughout all the mythologies and folklores was that there was always a lesson or warning presented in the stories. So I can understand the Indigenous People's fear of upsetting the spirits," said Aliah in the most agreeable and calm tone, which was another of the qualities that Afa found endearing about her. "I'm wondering about something regarding your work, Afa. Hearing you talk about this new boss, the environmental scientist and how obtuse he is, I always thought that Indigenous Peoples and environmentalists were on the same side?"

"I can see how you'd think that, but it's not always been the case. In fact, did you know most of the national parks that were created in the states, Latin America, and Africa were created by removing Indigenous Peoples who lived there? Those peoples were cohabitating with their environment sustainably. And they were removed so that environmentalists could visit those parks in their hiking shoes and see a so-called "pristine" environment. It's all a big fallacy. No environment has ever been pristine or untouched. Indigenous Peoples have always impacted their environments. In some ways visible, like the pyramids in Central

America, and in some ways invisible, such as crop rotation and permaculture. Based on this Western belief that natives lived in "wild" environments, the environmental elite removed Indigenous Peoples from their homes to keep the environment "wild"." Afa put air quotes around the word wild and finished her diatribe with a loud exhalation.

"Yes, girl, and that is why you are so good at your job. Because you learn about the communities you work with."

Afa looked sidelong at Aliah, hesitating. "I don't know, Aliah. I just…I feel like I missed it. I didn't even know about some of that folklore until I ran into this biologist on the way out of the Amazon and he told me. I could have missed it entirely. So, am I good at my job?"

"I think you are," Aliah started, but Afa held up her hand.

"No, girl, let me talk. This is bugging me." Aliah made a gesture to indicate Afa could go ahead, with a look of gentle concern. "I just don't know if I effectively explained the science, that I convinced them. And then when I realized I had missed learning about an entire worldview, a cultural and spiritual way of thinking, I wondered what else I missed, you know?" Aliah nodded, as if she understood. "Remember when you told me that one time that you felt broken?"

"Yes. But that was just a feeling, Afa. You're not broken. We break, we mend, we grow…"

"But that's the thing. Sometimes I have such a hard time admitting when I fail, admitting my doubts…it's like if I admit it or say it out loud, then everyone will know I'm just faking it. They'll see me for what I really am, which is an imposter."

"Afa, everyone has those little whispers of doubt, you know that, right? I've told you I struggle with that as well."

"But…you even told me yourself that when you feel like an imposter, it's fleeting, just in the moment. Deep down, you don't believe it." Afa felt a small lump in her throat, and she paused to swallow it down. "I believe mine."

Aliah grabbed Afa's hand and squeezed it, then let go, and Afa knew her friend understood her need for just a little space around this vulnerability. But in her eyes, there was so much compassion. Aliah said, "All those self-help books on your nightstand and your 88-year-old Jewish therapist haven't been able to lead you to the origin of that story?"

Afa blinked at Aliah, then scrunched up her face. "Well. Yeah. I guess."

"You're smart, Afa. You know what to do with this." Aliah leaned in, lowering her voice. "Your father may have emotionally beat the confidence out of you then, but I don't see him here now, do you? Stop questioning yourself. You are your own worst critic. You're not going to be perfect, Afa. But when you know more, you do better, and that is what makes you good at your job. You are an adult now. You have to show up as an adult, and not as the kid who wanted your dad to affirm you. You gotta affirm yourself now. You gotta believe you are good at your job. For the sake of your job."

Afa looked at her friend. This friend who had known Afa long enough and well enough to see past her bullshit, hear her out, call her on it when it mattered. Aliah was one of the only humans on the planet that Afa could open up to like this.

"Thanks, Aliah." Afa smiled, and again found she had to swallow a small lump down.

"Now, I will admit I also had a naive belief that indigenous lands were "wild," for lack of a better word." Aliah mimicked Afa's air quotes, returning to their earlier topic and resetting the emotional charge in the air. Again, Afa was grateful.

"Yes," Afa held up her hands in an affirmative gesture. "I can see why you thought that. It's the whole romanticized European construct of the primitive native that lived off the land, leaving it untouched. When I was researching the history of forests in national parks, which are mostly dried up now, I came across the case of the Batwa in Uganda in the 1990s. International environmental conservationists, backed by the World Bank, created reserves which overlapped with Batwa ancestral territory. When the park learned that the Batwa were hunting and eating silverback gorillas, which were recognized as a threatened species, also a main attraction for ecotourists from Europe and America, the Batwa were forcibly removed from the park. The Batwa denied killing the gorillas, claiming that other tribes had invaded and killed them. They said they felt kinship with the gorillas, but it was too late. Traditional western conservationists believed that the "wild" environment preserved as a park was incompatible with human communities. Even if they were hunting the gorillas, they probably knew how to sustainably hunt so that the species keeps proliferating. Environmentalists have kicked Indigenous Peoples out of state parks where they have always lived, so no. The global elite environmentalists and Indigenous Peoples – not always on the same side."

"Wow. It just sounds like environmentalists are not the do-gooders that I always thought they were. You are complicating my view of them," said Aliah.

"Yeah, it's complicated for sure. Indigenous Peoples are complex. Some of them want modern conveniences, and others don't. There is a true conflict of tradition versus modernization within indigenous populations. Corporations have swooped in and taken advantage of that tension, offering them schools, hospitals, roads, in exchange for extraction rights. And the tribe that I just met with is both modern and traditional. They have set up ecotourism and an international science station on their land.

And yet, they don't want the rain program because they think that the chemicals used in the rain production will harm their spirits."

"So what does that mean for your work? Your job? From what you are saying, are they going to reject the environmentalist agenda based on their folklore?"

"To be honest Aliah, I don't know. I just don't know. I think my job is to make folklore compatible with science."

"Geez. That sounds rough. That sounds like a dilemma that is centuries old. I mean, there are parts of this country that still consider evolution to be a theory that conflicts with the bible. There are schools in deep Christian parts of the country that don't teach evolution!"

"Yeah, right? And it's not just the tribe in Guyana I have to get agreement from. I need to see what happens with all the other tribes too. I need all the tribes to consent to the agreement to do an Amazon wide program."

Aliah sighed a deep sigh and looked down for a few seconds, very still, then popped her head up as if a cartoon lightbulb went off in her head. "Hey! It's not all bad, though, right? Tell me about all the other parts of your trip! Were you laying there while monkeys and shit were swooping up and down around you? On the jungle floor?" Aliah laughed, shifting direction and lightening the mood.

Afa grinned. "It wasn't a jungle floor! I was staying in an indigenous owned eco-tourism park that has a science station for researchers to study their environment. I slept in a room in a building!"

"But there was mosquito netting around the bed, right? And no windows? Just big holes in the sides? I could never sleep in the jungle!" Aliah cried. "Seriously though. Besides meetings, what do you do in the jungle?"

"Girl, hate to kill the image, but there are windows." Afa laughed. "There's so much to do there. The Indigenous Peoples lead tours of the patch, they take people in boats on the river. And you will be thrilled to know that, yes, they do have monkeys, Aliah." She grinned at her friend, who clasped her hands to her chest and swooned.

"I knew it! What kind did you see?"

"So many capuchin monkeys – the ones with the white faces. And spider monkeys. Those are the little, tiny ones. They like to come down on the branches and get close. The howlers are loud, and you can hear them early in the morning. Plus, all types of birds – super colorful ones. Toucans – you know the ones with the colorful beaks? And macaws…red and blue ones, oh, and even some crocodiles." Afa enjoyed describing the jungle to Aliah, watching her eyes light up with excitement the more she went on. "I wish you could come with me too."

"Nah, I'm cool here in NYC. My home is with the rats and the pigeons. The smell of rotten milk in the garbage can on the side of the street. I'm not really the Amazon jungle type. All those bugs make me want to take a shower in bleach. I would freak out if I got a parasite. And if something bit me, I would want to go straight to the ER!" Aliah mock shivered. "Seriously, though I love the work you do, Afa. You are out here in these streets, working for the climate, working for the people. You are an Americana tropicana now. It's… pretty dope," she sang these words, twisting her head side to side and raising her hands out of the water.

"Yeah, well...let's see how it goes." Afa gave a tight smile. She wanted to feel like she deserved Aliah's praise; to feel she could effectively marry the disparate worlds of science and folklore in such a way as to accomplish the task ahead of her. She shook her head and sighed, wanting to move on from the subject of her job for now. "So, anyway, enough about my work. What's up with you? I want to change the subject to something fun. Gimme the dirt on the dating scene. How is it?

"Nonexistent." Aliah gave her a dead eyed expression and then pushed out her full lips in a pout. "I mean, I really want to date someone, but I'm not meeting anyone that I connect to. I go on dates, and the vibe isn't there, you know? I've been thinking about my ex, Darius, and how I would just stare at him and think that he was such a beta. Both in the world, and in the bedroom."

"That's what you needed then. Nothing wrong with a beta."

"But I can't stand the idea of being with a beta again. It disgusts me. Now I see a beta as weak. I think this is a reaction to my beta ex and I don't want to be a prisoner of my past and just keep reacting, just swinging from one extreme to another. But I wonder, as an alpha woman who likes to run shit in my classroom, in my home, am I destined to be with a beta? Is that my future? Are all strong women leaders destined to be with betas or can two alphas be together? I want an alpha that pulls my hair in the bedroom, calls all the shots. I want an alpha outside the house, running things at his job, and I wonder if we can truly power share. Does an alpha man want a beta? Does it make their lives easier? In this day and age, we have deconstructed gender, yet there is this question of how power is divided between a couple. Most men today, probably, for primal and instinctual reasons, want to be the boss. Can we both be the boss? And share the power?"

"But Aliah, you really like being the boss in every area of your life. How are you going to share power?"

"Well, I want to reconcile the desire to be a powerful woman and the desire to be with a powerful man. Like Beyonce and Jay Z."

"You will find that powerful alpha man, your Jay Z. But at the same time, they're a power couple in the public eye, but who knows what their dynamics are in the home, in the bedroom? We don't know the details of their relationship. Can you think of any other alpha couples? I can't think of them right now – at least successful ones. Your thoughts are the musings of a powerful woman who knows what she wants, and a beta doesn't have to be your destiny if that's not what you want," assured Afa.

"Yeah. Maybe. But also, I just feel like I am running out of time." She paused and looked serious for a moment. She took a deep breath and said, "I'm wondering if I should freeze my eggs."

"Why are you in such a rush, Aliah? You're in your late 30s. You have plenty of time to meet someone."

"Not everyone is you, Afa. I want to meet someone and have a baby by 40. Even at 40, it's considered geriatric birth. I gotta meet someone ASAP."

"What does that mean, not everyone is you?"

"It just that you're not interested in a relationship, that's all."

"Who said that? I would welcome a relationship if it happened, but it's not my focus. You're focused because your biological clock is ticking. Mine is not. I figure I can always adopt. I don't need to give birth. I'm focused on my career. I'm not saying that you're not focused on your career, but I don't want to slow down right now. My job is super important to me, plus I have to travel for my job, and a relationship would tie me down," said Afa.

"Yeah, I know all that, Afa. But I think it's more than that."

"Like...how? What? You think I secretly want a relationship or something?" Afa bristled at what she thought Aliah was insinuating.

"No, no. I believe you mean all that. But I genuinely think the real reason that you're not focused on a relationship is because of what you experienced as a kid – watching your parents' dynamic. You observed your mom not having any control over her life. She had no "agency," as you would say, over the decisions your dad made about when to leave the island. You told me she wanted to leave earlier, but he didn't listen to her. You said you were one of the last to leave and things had gotten bad on the island. I think you don't want to be tied down by a man because of how your dad treated your mom," said Afa, her eyes softening as she spoke.

"Well, okay, Dr. Aliah." Afa smirked. "I mean, yeah. I guess there can be some truth to that. I don't want someone to affect how I make my decisions, of course. My independence matters to me, too. Which is a big part of why I don't want to be tied down right now. But," she sighed and gestured at the empty lane in front of them. "I think we've had enough analysis. Let's get in the pool."

They slipped into the tepid water and swam for the better part of an hour. Then, they showered while loudly talking about personal details of Aliah's last date. An older woman in the open shower stall next to them laughed when Aliah was describing her date's bad hygiene, which was another pet peeve for Aliah. The older woman tried to chime in over the shower stalls and said, "Sounds like my ex-husband," but Aliah was not having it.

"Excuse me, we are having a private conversation," Aliah said as her facial expression implied how dare you?!

"Excuse you! You are talking very loudly in a very public place!" replied the older woman.

Aliah shot back a look of death to the older woman and continued talking to Afa. This was an interaction that was very typical of New York City, thought Afa. She felt the line between public and private was always blurred because of the proximity to others all the time. It was the same in small restaurants where people were seated so closely to one another on small square tables, you could hear every word of the conversation of the couple next to you. On the island where she'd grown up, everyone had space. People were not on top of each other.

Back outside on the snow-covered streets, Aliah and Afa discussed which gallery to visit today. As was their routine, they'd finish their swim and go to a gallery. The sidewalk had just been shoveled, and the streets had a thin layer of powder on them as they walked to the Gagosian Gallery in the West 20s in Chelsea.

"So, you know that new boss?" Afa started, then she grimaced and said, "Ugh. Sorry. I'm talking about work again..."

"Ha, Afa. Its ok! You can talk to me about it. Tell me. What was your first impression?" Aliah nurtured the conversation along.

"Well, he definitely thinks he is doing god's work, which is almost arrogance. He's a very technical scientist, not much experience working directly with communities. People like that usually struggle with social and emotional aspects of a job. But honestly, I'm totally guessing. He could still surprise me," said Afa, as they entered the gallery. The women silently observed the art on the walls, roaming the halls together for the remainder of the afternoon. There was a new abstract exhibit, with lots of textures and deep thick layers of paint. The palette was soft and pastel with traces of gray hung on white walls with exposed brick. There was also an exhibit of figurative paintings, different male and female figures that looked like they were crashing into one another.

At one point, they stopped together in front of a painting of two men; one of the men had his fists out while the other man was down on the ground, but their expressions looked neutral.

"What do you think they are doing? Are they wrestling or fighting? Are they hurting each other, or having fun?" Aliah asked.

"I dunno. I guess...I guess I can't tell whether they are hurting one another. They could be goofing around. But I think... yes, I think they're fighting."

"I think they're wrestling, but having fun while they're at it," said Aliah. "How do you feel about it? Does it evoke any emotion in you?"

Afa glanced at her friend and smiled. She enjoyed when Ms. Aliah came to the surface; she was a natural teacher and couldn't help engaging those around her, inspiring them to take note of their surroundings.

"Well. Let's see," Afa sighed. "It makes me uncomfortable. I don't know what they're fighting about, and I don't know if the fight is fair. This one guy seems bigger than the other one," Afa pointed. "I don't like that it's unfair."

Aliah observed what Afa was pointing to. "Interesting. But maybe the smaller guy is smarter and faster and brings something to the table besides size. Maybe he is more clever."

"I don't know." Afa shook her head. "It's just a gut feeling. I don't like this painting."

"I think that is because you assume the bigger guy is going to win the fight," said Aliah.

"Don't they always?" responded Afa.

THREE

The sun baked the earth and cracked it open like a swollen, ripe peach. The thick canopy of trees above the Ecuadorian forest provided relief, and the sky darkened as the waxy, broad leaves brushed the car windows. Afa soon learned what the machete in the front seat was for, as her driver stopped the car every so often to chop at the matted plants and low hanging branches to allow the car to pass through. Dense thickets of bright tropical fruits hung low on the branches of the passing trees, and the rich hues of pinks and purples on the flowers were so vivid, they didn't seem real. The birds sounded like they were making their urgent calls to one another, and the smell of the rich, wet soil wafted up, so that her senses felt assaulted from every angle.

They drove past a small farm where bananas and coffee and yucca grew. She had read that this patch, Pastaza, was as big as New York City and the five boroughs, and it took four hours to drive from one side to the other. Located in the eastern jungle of Ecuador, these lands were inhabited exclusively by Indigenous Peoples.

Her presentation had gone very well. They'd been receptive to the science of climate modification, cloud seeding and the rain program as a whole, with none of them seeming to even flinch at the mention of the chemicals. Just like the Guyanese Macushi, they would give a final decision after discussing with their tribal communities.

Afa had been invited to stay in the village of the Pastaza patch after the meeting, and she knew that was a big honor, an invitation that was offered to a very select few. She felt generously welcomed. She'd read about Pastaza with great interest in her briefing. Two of the tribes sharing the province – the Shuar and Achuar – had once been fierce enemies. They opposed one another in the early 2000s when the government of Ecuador allowed oil companies into the Amazon to drill under ancestral tribal land. The oil companies had promised schools, hospitals, roads, and other much needed resources, along with a payment to the tribes. The two tribes went to war, with one tribe opposing drilling and the other wanting the benefits being offered by the oil companies. But the opposing tribe, with the help of pro-bono international lawyers, finally won the case in the Inter-American Court of Human Rights. In the end, no drilling was allowed, but as the rainforest began to shrink, both tribes were losing land. Facing mutual problems, the warring tribes had come together to peacefully share one of the two patches left in Ecuador.

On the long drive to the village after the presentation, surrounded by the sleeping tribal elders, Afa thought about the land right outside the patch. She had moved through an ocean of dead trees, ravaged by heat, fires, and deforestation. She had the privilege to visit what was left of the Amazon, patches of humid rainforest in an otherwise dry wasteland of dead trees. And the patches were getting smaller, fraying at the edges. When the forest began to dry up, many of the Indigenous Peoples had moved to major cities in Guyana, Brazil, Venezuela and Ecuador, but many had stayed behind. They were losing their land, thought Afa, but at least they still had a country. When she left her island and it was submerged in the ocean, her passport became meaningless.

Hours later, they arrived at a village amongst a tall canopy of trees with a humid understory of plants all around. The colors were as loud as the sounds of birds she had not heard before, the howl of monkeys so loud, she felt they must be only three feet away but learned they could be as far as three miles away. As she got out of the car, she was guided by the elder Luiz who introduced some of the other elders and arranged for her luggage to be taken to her room. Luiz seemed at once to be both patient and impatient. He sounded irritated at the thought of setting up Afa in her accommodation and delegated that very quickly, but then he turned to Afa.

"Come Afa, we have some chicha for you to drink. We offer it to our special guests when they arrive," said Luiz.

"I have always wanted to try chicha. How do you make it?" Chicha was an alcohol made from yucca that had been brewed by communities in the Amazon for thousands of years.

"The older women chew the yucca root and spit it out. The mastication with their saliva creates the fermentation of the Chicha." Luiz's eyes glinted at her mischievously.

Afa tried to conceal her reaction. Obviously, there was nothing too appealing about the idea of drinking something that other people had chewed up with their mouths and spit out, but she would never turn down a traditional drink from Indigenous Peoples who had been so gracious to host her. So instead, she just nodded and smiled, trying to convey enthusiasm she didn't feel. "Wow. I had no idea..."

Luiz gave her a small smile, then led her to a circular structure with a roof made of dried plant materials. This is where other people in the tribe were waiting to greet her. She introduced herself to a mix of women and men, many with black markings on their faces. She sat down with the others as a circular bowl was quietly being passed around to share. The bowl looked like a carved-out coconut or seed shell. When the bowl came to her, she didn't hesitate. She took the bowl, and willing herself to disregard the idea of how it was made, she took a large sip of chicha. A fermented taste hit her palate, followed by a slightly sweet aftertaste. She pulled the circular shell away from her mouth and smiled, then noticed small brown bits floating in the chicha.

"What are the brown bits?" she asked Luiz, pointing into the liquid.

"Oh yes, that is the old women's teeth that fell out from all the chewing," said Luiz.

Afa's eyes grew large, and she resisted the urge to cover her mouth and spit out the chicha. She tried to arrange her face into a smile so that they couldn't sense how grossed out she was, dearly hoping not to offend them.

When Luiz saw her face, he burst out laughing, and then everyone sitting in the circle began to laugh and Luiz said, "I am just teasing you! Those are not women's teeth! It is not spit!"

Relief flooded through Afa, and she began to laugh herself, shaking her head. Honestly, she wanted so much to gain their trust, if they had been teeth, she probably would have just gone ahead and endured it.

As the chicha went around the circle, people from the two different tribes of the Pastaza patch spoke to Afa about their lives. They introduced themselves in English, though most of them spoke four languages – English, Spanish, Shuar, and Kichwa.

Afa could hear that they were well informed of their international human rights, and Luiz's daughter had left the tribe for the capital city of Quito, earned her law degree, then moved back to the indigenous land. They knew of international politics, and they had close alliances with organizations all over the world that were helping them to preserve their lifestyle. The oil companies were still in the Amazon, and had been since the 1970s. The fact that the Amazon was drying up meant that governments tried to argue that no Indigenous Peoples lived there anymore so drilling could begin or in some cases, commence. But this patch she was staying on was protected, fought for by international lawyers, safe for now.

After the chica went around the circle, Luiz turned to Afa. "Would you like a tour of some of the medicinal plants that grow in our forest?"

Delighted, Afa nodded vigorously. "I would like that very much!" Luiz and Afa walked slowly towards a clearing just past the circular structures where the tribe slept and ate. They entered the dark forest, canopied with thick trees that blocked out most of the light. Here and there, narrow rays of sunshine streamed through the leaves, gingerly touching the plants growing on the ground at the base of the trees.

"There are three levels to our forest," said Luiz. "There are the top layers where the monkeys and birds reside, there are the middle layers where the fruits are, and there are the bottom layers where the plants and trunks of the trees give us medicine."

As he pointed out different plants and a few tree trunks and leaves, he told Afa about the healer who lived amongst them, Franco.

"Franco has taught me many things; you'll get to meet him this evening. Do you know the difference between medicine and poison?" Luiz asked.

"No," responded Afa.

"Dosage," responded Luiz. "Dosage is the difference between medicine and poison." The tour continued as Afa felt different plants, smoothed her hand over the trunks of trees, picked leaves to smell them, and felt the moist understory of the forest beneath her feet as Luiz explained what different qualities and properties the plants had. They slowly walked back to the circular structure.

Afa sat amongst the community as dinner of fragrant fish soup and bananas was being served. It was delicious. The fish broth tasted earthy. The bananas were sweet and silky tasting, not mealy like bananas she ate in New York. As she ate, she heard and saw the colorful birds perched on trees outside the circular structure that was providing them shelter from the soft light rain that fell every hour or so. It was the intact water cycle that so much of the Amazon had lost. The water on the patch was still being absorbed by the trees, evaporating and condensing in the air and falling again on the forest. The patches were centered around the last rivers that still flowed with water, eventually emptying into the ocean. It was beautiful and she couldn't help but contrast it to the dry savannah they had to drive through to get to the patch. She wished that the whole Amazon felt how this patch felt. It was so humid, her pores on her face had opened as if she was in a sauna; such a welcome feeling after visiting so many dry parts of the world. But she knew that the patch would not last long as the savannah kept spreading. The indigenous communities barely had enough animals to hunt. There were fewer fish in the rivers. There was less of everything; fruits, medicine, and the raw materials to build their homes in the forests. And of course, every year the rainfall diminished, meaning less fresh water to keep the forest moist. Afa was starting to wriggle in place, attempting to rearrange her feet. Would she be able to convey the urgency of this program to the people who called the forest home? Afa wasn't sure, but she knew she had to try for the sake of the people and the future of this beautiful spot on the planet.

Afa wondered if it would be disrespectful to bring up business, to ask about their thoughts on bringing the United Nations rain program to the Amazon, but she held back. After dinner had been cleaned up, Luiz told stories of his daughter living in Quito, proudly describing her success at law school. Afa listened intently, truly interested. She waited for a natural pause in the conversation and then spoke.

"May I ask you all what you think about the rain program coming to this region?" she asked nervously to the whole group in the circle, hoping her timing didn't go against custom.

"We are very excited about the program!" said Luiz's wife Cristina. "We are happy that more rain is coming and that our patch will grow at its borders. We believe that the forest can come back."

"That is wonderful to hear!" said Afa, a weight lifting off her chest, her shoulders softening with relief.

"We are very excited about the science of climate modification; the whole tribe is excited. We consider ourselves a very worldly patch. We have many professionals here, and we do not want to be left out of the modern advances of science and technology. This is truly a breakthrough, and we feel so fortunate to be a part of such innovation!"

"I'm thrilled that you want to be a part of this program. But I'm curious…are you at all hesitant about the chemicals?" Afa wondered if she should even bring that up and ruin the mood.

"No, we have done our research when we were first offered to be a part of this program, before you came. We know the chemicals are harmless. It's a small price to pay. Our biggest concern is making sure that we are involved in the process. We understand the United Nations is responsible for this program.

But there are many of us indigenous nations within the forest in different countries that will be affected, but COICA must be involved," said Cristina.

COICA was an acronym for "Coordinadora de las Organizaciones Indígenas de la Cuenca Amazónica" and was the umbrella organization of all the different Indigenous Peoples' tribes in the Amazon. COICA advocated for and coordinated their political interests on an international level and spanned through eight Amazon countries. She had exchanged emails with members of COICA and had a video call scheduled for when she returned to New York City.

"I am meeting with COICA leaders next week," said Afa.

"If you can guarantee our involvement, when the rains will come, how often, how much – we know the rhythm of the forest and our voices must be heard. But we are looking forward to the program," Cristina said excitedly. The others began to voice their enthusiasm for the program, too, and they told many stories of what could happen if the rain came back to other regions where other tribes had lived. What It could mean for the landscapes of dead tree stumps that were found a few hours outside their patch.

Afa, encouraged by the enthusiasm, grinned broadly. "I'm so happy you're on board!"

"But we have heard that tribes in Guyana do not want the program," said another woman named Sylvia. There was a long deafening pause in the conversation.

Afa had anticipated that tribes communicated with one another through COICA and she thought they might be aware of the Macushi resistance, but she'd come prepared for that and rolled out what she'd practiced in response.

"The most important thing is for the Indigenous Peoples to have agency over the decisions that will affect them." She paused and looked at all their faces, then turned to Sylvia.

"Do you know what their reasons are for their concerns?" asked Afa. She was curious if they would repeat the folklore that Vishnu had told her about.

"I think that they must tell you themselves, they must represent themselves. I do not want to speak on their behalf. But I will say that the Pastaza patch is governed by tribes that do not follow stories and myths that do not serve us in a forest that is dying. We choose science to improve our lives, not stories. The stories were meant for a different time that does not apply now," said Sylvia.

"Yes, of course," said Afa quickly, realizing it was inappropriate for her to have asked Sylvia about the Guyanese tribe. She was intrigued by the contrast of their beliefs. But she also realized it was not right for her to ask one patch about the customs or beliefs of another patch. The others sitting in the circle began to ask her more specific questions about the rain program and she answered as many as she could. Afa explained again that while she wasn't a scientist herself, she knew only of the procedures to gather consent, but she assured them she would find out as much as she could upon her return to New York.

After thirty minutes of questions, Afa felt a dull pain in her abdomen. "My apologies, but I have a stomachache, and I think I should lie down. Where will I be resting tonight?" Luiz showed her to her room and showed her where the others were sleeping in case she needed something.

Afa was so grateful for her accommodations and as she lay down, her stomach suddenly began to gurgle. The rumble was dull at first but then turned violent. Oh no, she thought, quickly bolting from her room across a stretch of forest clearing to go to the bathroom. No sooner had she returned to her room and lay back down when her stomach seized in pain once more, and she jumped up and flew across the clearing again. Her skin had grown clammy, and sweat was beading on her lip and brow. She tried to settle back in her bed, but tossed and turned, writhing in pain. In one hour, she had dashed across the dark forest trail to use the toilet four times. After her fifth trip to the toilet, weakened and shaky, she knew it was time to find Luiz. She walked over to his structure and called out to him.

"Luiz?"

"Yes?"

"I have gone to the bathroom several times in the last hour. Is there anything I can do?" she asked.

"I will go find Franco, our healer. Go back to where you are staying and Franco will come to you," said Luiz.

Afa walked back, lay down and continued to feel small spasms in her stomach, but after all those visits to the toilet, she didn't see how there could be anything left. After about ten minutes, a tall, skinny older man entered her room and slowly knelt with some difficulty next to her. "Hello, Miss Afa. I am Franco. I will try to help you feel better. I understand you keep using the bathroom. Is there anything else?" he asked

"Yes, I have some sharp pains in my stomach, and I feel clammy."

"Ok, I will be right back," he said.

As Afa lay there, she considered that she was a six-hour drive from the nearest large hospital, and she prayed Franco could help – she had to place her faith in him. Perhaps it was the food that did not agree with her stomach, but for fear of offending her hosts, there was no way she would say that out loud. Franco returned to her room within five minutes, but felt more like fifteen.

"I am going to ask you to drink water with a few drops of something we call Sangre de Drago. It is a sap of a tree that grows here in the Amazon. It is like a liquid bandage, and it is going to help with your diarrhea. The next thing I am going to do is rub some snake venom on your stomach for the spasms. It will help you with the pain," said Franco. He asked permission with his eyes to begin to rub the venom on her stomach. Afa nodded and he began. As he finished, he said, "I will be the first person you see when you wake up, now go to sleep."

Afa had a restful night despite some slight pain in her abdomen. She had stopped going to the bathroom and was able to lay still and sleep for about seven hours. When she opened her eyes, Franco's face was the first thing she saw, as promised, his furrowed brow and squinted eyes staring intently at her.

"How are you feeling?" Franco asked.

"Much better, Franco. Thank you."

"What did you dream about?" asked Franco.

Afa was surprised that she remembered her dream, which was not usual for her. "I dreamt that I was in a pool with my friend Aliah, and the water was rising, and I wanted to yell at her to get out of the pool, but every time I opened my mouth, no sound would come out," said Afa.

"Yes," said Franco. "I have spent most of my life interpreting dreams here on the patch, and I can see from your dream that you care very much, and you want to help others. Your work here concerns you deeply. Please let me know if there is any other discomfort today." He quietly stood back up, with a slight groan over his creaky joints, and walked out.

Afa's driver would arrive in a few hours to take her back to Quito. She'd hoped to spend the day with her hosts, but the episode last night had left her with a lack of energy. She opted to stay in bed and rest, but her mind swirled with the details of her next trip to the Amazon to see the Guyanese tribes. The stakes were high for her to convince them to consent to the program, and it would be necessary for her to show what the rain program would be able to achieve. She was hoping that she could balance their perspectives on their folklore and their concerns about the spirits as well as educate them on what their agreement would entail. In her mind she replayed the previous presentation she had made in Guyana, her guts squirming now not with the upset from the night before, but fear and trepidation over the task ahead. She seemed to have been effective in Ecuador, and that made her feel good. But was it her effectiveness, or was it more that the Pastaza tribes had an outlook of positivity towards the program even before she arrived?

For now, she was comforted by the thought of going home to New York and the UN building, with its cement walls and large glass windows that looked out to the East River. It was still winter in New York which meant the sun set before she left the building to go home. On those nights, the sky was a soft pink, streaked with peach and lavender.

When Afa's car arrived, Franco called out to her to tell her to tell her the driver was here. Afa left her room and saw many members of the tribe standing by the car. She was moved at this sight. They had assembled to see her off, and she saw many people she recognized, Franco, Luiz, the women she spoke to about the rain agreement, the people she drank chicha with.

"Thank you so much everyone for being such good hosts, I truly had a great time," said Afa.

"Even though you had a stomachache? I bet that was your first time having snake venom on your stomach!" said Franco with a twinkle in his eyes.

"Yes, even with the stomachache, and you are right, I have never had snake venom on my body."

"Don't try that at home. Those of us who live in the forest know which venoms can heal and which ones can kill!" said Franco, giving Afa a wink, and she smiled back at him.

"Afa, thank you for bringing us the rain program. We are excited to be a part of it, and we are so happy that there is a possibility that rain can come back to the whole Amazon. We know our patch is getting smaller, and we are so happy that the international community cares about our home, our culture, our people," said Luiz.

Afa felt so warm inside hearing Luiz's words. She had done her job well, and she believed that she had communicated to them in a way that they had heard her. She was relieved and an inner calm spread through her chest. Her shoulders seemed to soften and drop an inch.

One of the younger women with beautiful, elaborate black markings on her face approached Afa.

"We have made you a gift of bananas from our trees for you to eat on your travels back to Quito," she said, as she held out her hand to reveal bright yellow small bananas wrapped in a large green leaf.

"Thank you so much." Afa bowed her head. "What is your name?"

"My name is Nina," the young woman replied.

"Watch out for Nina," said Luiz in a booming, proud voice. "She will grow up to be one of our leaders. She has a lot of vision; she is a seer. She sees things she has not been told."

Afa wanted to ask more about Nina and her abilities, but this was not the time. “Thank you all so much for all that you have given me. Your time, your amazing food, your medicine. I loved learning about your forest.” Afa graciously bowed her head as she said, “Thank you, Thank you. I will speak to you again.” As they said goodbye in a chorus of different voices, Afa stepped into her car.

“Looks like they liked your visit. You are lucky, not many people are invited to stay with them,” said the driver, smiling back at her.

“I know, I am very lucky,” Afa said as a smile spread between her cheeks. They began the drive out of the patch, a sea of tall trees and understory and soft underbrush, until they reached the end of the patch where the dead trees began. The decay had spread across the dry rainforest as far as the eye could see.

They finally reached Quito. At the airport, Afa ordered a tangy, fragrant passionfruit juice that quenched her thirst ahead of the long journey.

Afa had an eight-hour layover in Miami before she got to New York City, and Kwame had scheduled a check-in with her during that window of time. After departing the plane in Miami, using the restroom and getting a snack, she sat waiting for his call. Finally, a few minutes past their agreed upon time, the phone rang as she sat in a quiet corner of the airport without much foot traffic.

“Hello! Afa! Sorry I’m late. I was just on a personal call that went on a bit long.”

“Oh…is everything ok?”

“Well, yes. I mean not really,” Kwame paused. “I don’t know if I should talk about it. Not really related to work. But I’m distracted.” He paused again, then Afa heard him take a shuddering breath. "Sorry, Afa. It's my dad. He was just diagnosed with Parkinson's, and he won't take the prescribed medication because he is worried about side effects. My mother and I are trying to convince him to go on the meds. It's unbelievable that I have to convince a retired psychiatric nurse to take medication."

"Oh, Kwame, I'm sorry to hear that, I really am," said Afa, a bit startled by how much he was sharing with her.

"Thanks. Yeah, it's just that I'm pretty involved in my dad's care. My dad took really good care of us when we were kids. He always reminds me of this one time when I was eight years old and in school we had to write about our dads. I wrote, 'I hate it when my mum is sick because my dad cooks and his food is bad.' He is still sore about that." Kwame chuckled softly.

"Oh, that's funny," Afa laughed, touched by Kwame's candor.

"Anyways, that's what's going on with me. Sorry to dump a load of personal stuff into our chat. But since I did, what about you? Are you close to your father? I mean, do you see him often?" Here Kwame paused, but Afa remained silent. "Afa, I'm sorry. I hope this conversation isn't too personal, I was just trying to relate," said Kwame in a rush.

"Oh, no. It's fine. It's not too personal, but thanks for checking. But my dad and me? We are a bit distant." Afa didn't want to say more, but Kwame's silence lingered, and she found herself going on. "I guess you could say my dad is stubborn. He wouldn't factor in other people's perspectives, especially my mom's, regarding the rising sea. When the island began to be submerged, people began to seek environmental refugee status all over the world. The first wave went to Australia. My mom was a part of the first wave of people who wanted to leave. She saw what was happening all around us. The sea water was eroding the pipes under the ground and the island infrastructure started to break down, salinization was happening. And that is when she felt it was time to go. But when she told my dad, it was like he didn't hear her. He told us that we would be one of the last families to leave the island because it was our home. He waited till the absolute end, when the island was almost completely under water."

"Wow, Afa. I had no idea. That must have been scary for you as a kid."

"I was just old enough to realize. My mom was so scared for us kids; she worried that we wouldn't get refugee status in time before the island became submerged. Finally, with so little time to spare, we got refugee status in the U.S., and we moved to New York. I was 12 years old. I just remember so many nights when my parents would argue, my mom begging to leave the island before it was too late, and my father just would not hear her. He didn't even stop to consider her concerns, her reasons for wanting to leave the island."

"That sounds so difficult, Afa." Kwame said, and Afa thought the empathy in his voice sounded genuine. "What did your father do for work?"

"He was a teacher at the school. In the end, there were barely any families or kids left to teach. Then they stopped paying his salary when there were so few kids, and we lived on our savings and help from my aunt and uncle until we got off the island."

"I see. So that affects your relationship now?"

"Yes." Afa suddenly realized how much she had shared despite herself, and a pang spread across her chest. She stopped talking, wondering what the price might be for making yourself so vulnerable to your boss.

"I hear you. Our fathers are the original authority figures in our lives. And how they parent us affects us for years. Unfortunately, and I suppose fortunately as well." Kwame waited a beat, then said, "Anyways, moving on, let's talk about Ecuador! How did it go?"

Afa was relieved he was changing the subject back to business. "It went great! I think they are going to sign the agreement. I'm just waiting for them to give their formal consent. Any day now, I think they will be signatories."

"That is excellent. Excellent news! I am sure Guyana will do the same," Kwame was unable to hide his enthusiasm.

Afa wanted to tell Kwame about how the Macushi worldview impacted their decision to possibly refuse the program. But giving him the impression that she couldn't handle the task at hand wasn't something she wanted to do, especially not after having just shared such personal details with him about her own upbringing. What if he got the idea that she was giving up before she began? No, she needed him to believe she could handle anything that came her way.

"Well, with Guyana, there was some resistance, like I mentioned in my notes. But I am hoping that I can convey the benefits in a way that outweighs their fears about the chemicals. We will see..." Afa cast her mind about, wanting to change the subject. She wasn't ready to discuss what she had learned from Vishnu about the folklore. In truth, she was worried that she wouldn't be effective with the tribes, but she would hold that close to the chest for now. That conversation could happen after she met with the Macushi in the Guyanese patch for the second time.

Instead, she pivoted back to their earlier conversation. "What are you going to do about your dad?"

"I don't know," Kwame said, taking the bait. "To be honest, it makes me think a lot about life and death. My father is very spiritual and believes there is life for the spirit after death. But my mother is not. She fears death, she believes it's the end. I think my beliefs are somewhere in between my parents"

"Ah, that makes sense. We get our worldviews from our parents, even when they don't speak them out loud," said Afa. She remembered that from a developmental psychology course she took at NYU. And, she noted, bringing the topic of worldview and spirituality into the conversation as it applied to his own life might help down the line.

"Yes. We do. Hey, thanks for listening and sharing, Afa. I think the qualities you have shown in this conversation tell me that you can gain the trust of the Macushi. I would suggest you engage with them deeper, more intimately, and participate in their ecotourism program next time you go. Even before you start to discuss the agreement."

Afa was very surprised that someone she had pegged as so strictly a scientist and a technician was talking to her about going deeper to gain trust. Perhaps she had read him wrong, been too quick to judge and, possibly, had judged him too harshly. Perhaps he would be empathetic towards the Macushi, she thought. Or maybe he wouldn't be, but she would deal with that all after her second visit to Guyana.

"Thank you, Kwame. I hope I can gain their trust too," said Afa.

"Ok. We will talk later. Have a great flight to New York. See you in the office."

Afa disconnected the call and sat back, considering her own worldviews and how they had shaped her.

FOUR

It was the second time that she had spent nearly the whole day driving from Georgetown to the Macushi patch through a dusty savannah, moving towards a tropical forest. Finally, eight hours later, with the driver's and her own belly full of food from the outpost, they were in the patch. The road was muddy, damp red earth winding through the matted trees. There was a small river crossing via a single small ferry boat to reach the eco-tourism compound and science station. She inhaled the natural earthy perfume and drank with her eyes the familiar sights of the rainforest. The trees were so tall, with the familiar speckles of light shining through the canopy. The river was flowing and there was a light humid rain every few minutes. She loved the sounds of the jungle; monkeys howling, birds calling, the rattle of tree branches, the rushing of the water in the river. As Afa stepped out of the van, across a clearing in the forest, she saw a woman with a steady gait approaching the car. She had such a strong sense of presence, Afa thought this must be Marcia.

"Hello," the woman called as she drew closer. "I'm Marcia. Welcome again to the Macushi Patch." She smiled warmly and extended her hand out to Afa. "I wasn't able to meet you last time, but I know you'll be here longer on this visit. I've heard that you will take part in our boat tours, our botanical tours, our guided walks in the forest – all part of our ecotourism program we offer here. I am very pleased."

"Hello Marcia, it's so nice to meet you," Afa said as she took Marcia's hand. "Thank you for having me as your guest."

"Of course. We have much respect for the work of the United Nations, as the Macushi have had visitors from all over the world. We understand the value of international cooperation, and we are happy to have you," said Marcia.

Marcia was strong and calm and serene, yet full of energy. Afa felt compelled to thank her more than once, and she had to stop herself from gushing. "Come, let me show you around our compound. I know last time you were here, you came and left on the same day and didn't get to see much. This time we want to show you the success of the Macushi enterprises. Leave your luggage for a moment and walk with me to the clearing where you can see everything." Marcia led Afa away from the red dirt road through the tall trees down a narrow trail that opened to a large, vast clearing. "That large circular structure is our main building here. We serve all our meals there to the eco-tourism guests and scientists, the students, the diplomats – people like yourself. At the base of the building is also where we have all our large meeting rooms for groups, as well as the clinic where you will go if you need first aid or health advice." Marcia continued walking along the red dirt path, pointing to a structure in the distance. "In this direction, can you see the long skinny building?"

"Yes," said Afa, in awe of the beauty and how the buildings blended in with the surrounding forest.

"That is our science station. That is where you will be sleeping. Our ecotourist bungalows were all booked."

"Oh, I see. I'm sure I'll be very comfortable. Thank you," said Afa again. For some reason, Marcia made her a touch nervous with the long pauses in her speech, even though she was gracious and friendly, showing Afa around and taking the time to make Afa feel comfortable. Marcia smiled kindly, pointing towards the river. "And off to the side near the river are the small bungalows I mentioned for our ecotourism guests. Each room has a view of the river and has a hammock strung outside."

"They look gorgeous. Maybe next time I come I can stay in one of the bungalows."

Marcia smiled and nodded.

"They are very nice. Rustic, but comfortable. As for the inhabitants of the patch, the Macushi, we all live deeper in the forest. You cannot see our houses from the compound clearing."

"Oh, I'm sure it's important to have a degree of separation from all the eco-tourism activity and the science center," said Afa.

It was hard not to be in awe of Marcia. She radiated natural confidence and competence, qualities that Afa admired in others. She had made such an impression already and they had not even begun to discuss anything besides the layout of the compound.

"There are a few rules for ecotourists here," said Marcia, transitioning into a more businesslike tone. "Number one – no swimming, as there are a few crocodiles left in the river. There are also piranhas, although we wish there were more. We have lost fish as the patch is getting smaller. But no swimming, ok?"

"Yes, understood." responded Afa. "All you had to say was piranha."

Marcia smirked and raised an eyebrow before continuing. "Number two – no walking on the trails without a guide. There are many snakes here on the patch and though we have many of the anti-venoms in the clinic, if you get bitten, there needs to be a local guide with you to identify the snake so we can use the right anti-venom, ok?"

"Got it. And what a relief that you have anti-venoms," said Afa.

"We have an anti-venom for almost all snakes except the Coral snake. We have an expression here, 'If the coral snake bites you, you don't go by the hospital; you go by the priest, because if he gets you, he really gets you.' The Coral snake destroys your blood and your nervous system. But they are not in the clearing. Here you will not find the coral snake, they are not usually along the wide trails either. They are found in other parts of the patch, and there are not that many, so it's not one you need to worry about. But you definitely must have a guide with you when you walk the trails," Marcia stressed.

Even knowing encountering the deadliest of the snakes was unlikely, Afa felt a chill down her spine. It made the New Yorker in her want to stay out of the deep forest and spend all her time in the clearing, in the main part of the compound. She shuddered at the thought of even seeing a snake, let alone knowing that it could kill her.

"Oh – she looks scared now!" a voice rang out behind them, laughing loudly. Afa and Marcia turned to find Noel approaching them, with a flowing yellow dress that moved as she walked. She still had the same slicked back hair and hoop earrings, but this time she looked more Georgetown than she did jungle. Afa loved that she had a PhD in science but had sensuality oozing from every swing of her hip. She had not relegated herself to being just a brain. She lived in her body, and lived there well. "Hello Ms. Afa! How lovely to see you again. You are really getting to know the patch now!"

Afa grinned and clasped Noel's outstretched hand. "Noel! Great to see you too. Yes, Marcia is just telling me about crocodiles and piranhas and snakes!" Afa laughed.

"I don't mean to scare you, we just ask that you follow these rules for your own safety," said Marcia calmly.

"Absolutely, Marcia, I will follow all the rules. I am nothing if not a rule-follower, particularly when life is at risk," Afa, ever the obedient student, nodded emphatically and held up her hand in oath, wondering if that American gesture translated to Macushi culture.

Marcia grinned. "Perfect. Now, as I have a meeting to get to, Noel will take you from here. Welcome again." She bowed her head slightly and turned to go.

"Thank you again, Marcia. I look forward to speaking to you more," Afa said, and then cringed. Had she really just said thank you again? How many times had she said it? How could she make herself stop?

Noel gestured towards the outcrop of buildings near the Science Station. "Come, I will show you where you will be sleeping. And if you need me, you should know I also sleep in the scientist's station. The rooms are small and there are communal bathrooms and showers. There are always frogs in the shower, but don't worry about them. They're harmless, so you can just shoo them away."

"Frogs are better than snakes!" Afa said with her eyes wide.

Noel laughed. "You should have seen your face when Marcia was describing snakes. You were white as a ghost!"

Noel's effervescent personality appeared to Afa to be a mix of Georgetown and the jungle. She could rough it out in the rain and muddy forest floor, but she was still polished. She was wearing full makeup in the Amazon – red lipstick, a full set of lashes – smiling large and laughing easily. Afa couldn't help but be drawn to her energy, to want to know more about this enigmatic woman. As they toured the science station, they spoke about world politics. Noel knew about the situation in so many different countries, their leaders and their political struggles. Noel was also approachable, whereas Marcia seemed to be floating on another plane. Afa respected both women, but they were so different. Afa was relieved that she was going to be sleeping near Noel so she could ask her all her questions. She seemed full of answers.

The room Noel showed Afa was small, with cement walls and floors and mosquito nets everywhere. Just as her friend Aliah had imagined. Afa put down her bags and stretched; it had been a long day and she was feeling the journey.

"You must be exhausted from your journey. Let me show you the bathroom and shower," said Noel. They walked over and just as had been predicted, there were two large frogs in the shower, staring back at her.

"I've got to run now, but you've got the frogs to keep you company." Noel laughed her warm, hearty laugh. "Get some rest. I'll see you a bit later," said Noel. Afa smiled and watched as Noel turned and walked away, her backside bouncing with every step, her voluptuous body moving slowly and gracefully through the forest like a thick jaguar.

***

The next morning, every hour or so there had been a light rain. As Afa stood outside her room, she inhaled the clean wet air and took in the surreal feeling of the rainforest. The ground was a muddy brick red and where there was vegetation, bright green popped. The pinkish purple tropical flowers that hung like vines from the branches of trees and the sound of the capuchin monkey's laughter just overhead delighted her.

The rumble of the large yellow van approaching broke her from her jungle musings. The van came to a stop near the science station and the Macushi eco-tourism employees began to pile out. She just stared at them, all working so diligently and quickly, carrying boxes back to the circular building. She was mesmerized that the Macushi had found a way to capitalize off the beauty of the forest, bringing tourists to the compound against the backdrop of the many pantone shades of lush green in the heavily layered forest. Noel, with her signature red lipstick, big hoop earrings and a dress that hugged her hourglass shape, was hard to miss as she emerged from her room.

"Hello, Ms. Afa! Good morning! Once you get settled after breakfast, I booked us a boat tour of this part of the river. I thought I could join you and we could talk a bit more, orient you to the patch," said Noel.

"Wonderful! That sounds fun! I'm thrilled to be in the rainforest patch again. Every time I stay in a wet jungle I'm disoriented by how loud it is. And that river! It's so wide and flowing, it's easy to forget there are still rivers so full and high in the Amazon. The whole environment here feels psychedelic, like my consciousness is being expanded," Afa said as she pulled out a large bottle of insect repellant from her pocket and started to generously apply it on her face, arms and legs.

"Looks like you brought enough insect repellant for a small family!" said Noel with a chuckle. "That won't help you here."

"Well, I've read about all the different insects that are thriving on the forest floor. I don't want to get any bites." Afa was suddenly conscious of how much she was acting like a tourist. She didn't feel as tough as Noel.

After the breakfast of papaya, fish, and toast, with coconut water and weak coffee, Noel returned. "Our guide for the boat trip will be Johnny, our very best. If you're finished with your breakfast, let's go down and meet him at the riverbank," Noel said with a spirited tone. Afa noted that Johnny seemed to be a common name in Guyana, even amongst the Macushi, reflecting the history of British colonialism of Guyana.

As Noel and Afa climbed into the boat, they greeted Johnny and thanked him for being their guide. He revved the motor of the boat and gestured for them to take a seat. They set out on a wide part of the river until they reached a small tributary where the water was calm and still. There were large pastel colored lily pads floating on the surface with little, teeny spikes. Afa reached to pick one up and quickly noted that the spikes poked her fingers. Across the river, mangroves grew tall with their large, exposed buttress roots holding the trees down in the river like gigantic claws sinking into mud.

"When do you have your meetings with the tribe?" asked Noel.

"Either tomorrow or the day after. Marcia said she and the elders are trying to coordinate a time." Afa didn't want to talk about the meetings in front of Johnny, and she was surprised Noel brought it up. It didn't seem appropriate to discuss in front of a community member, given the sensitive nature of their resistance to it. Perhaps Noel didn't know about their apprehension... Afa thought it was best to redirect the conversation. "Hey, the food here is fantastic! The bananas have such a strong taste and are so sweet, and the cooked meals are so delicious. I love that we all eat together – scientists, tourists, and the indigenous staff. It's a great arrangement. Oh, and I love that there is a clinic on site, and even a helicopter pad! They really have everything here."

"Yes, we really do. I live both in the jungle and Georgetown and I am so happy to have this arrangement. I couldn't imagine anything else." They spent another hour on the boat engaging with Johnny, hearing about all the different parts of the river and the different species of fish and birds that were specific to that patch. He was very knowledgeable as he pointed out the different species of fish: Arapaima, Piranha, Arowana, and Payara fish – which was also nicknamed 'the vampire fish.' Johnny kept saying that the vampire fish did not actually suck blood, but had two very large fangs, seemingly concerned that Afa might be fearful of the fish. Afa imagined huge fangs and shuddered, and she no longer wanted to bend over the edge of the boat and see the fish, but she laughed at Johnny's overly concerned expression and assured him she understood the fish was not actually a vampire.

"How would you like to hold a baby crocodile?" Johnny questioned, his eyes glistening with mischief. Suddenly, he yanked a baby crocodile out of the water and held it in such a way that it could not open and shut his mouth. "He won't hurt you as long as I am holding onto him."

Afa thought, No way in hell am I touching that thing, when Noel exclaimed, "I'll hold him!" with such enthusiasm, Afa gasped. She moved away from the tiny crocodile as Noel reached out to take hold of it. Wow, this is some real jungle behavior right here.

"Looks like you are not too fond of the idea of holding a baby crocodile," said Johhny to Afa. "Scared?"

"I think it's natural to be scared of something that can eat you!" retorted Afa.

Johnny threw his head back and laughed a deep, hearty laugh. "Afa! It's a baby! It can't eat you."

Afa, enamored with how playful he was, wondered what a younger man like Johnny's take was on the local folklore.

"Hey Johnny, what do you know about the Water Mamas?"

Johhny shrugged. "I've heard about them, sure. Mostly from our elders. But I don't know much. I kind of know them as mermaids who can take our people under the water, or hurt us, if we don't listen to what they want."

"And what do the Water Mamas want?" asked Afa

"I don't know. I don't really believe in it to be honest. I don't believe there are mermaids in there. It's just something our elders talk about, like a story they tell us."

Afa nodded. "I see. Do you think they believe it?"

Johnny shrugged again. "I think some of them do. Some of them just tell stories, but a lot of 'em believe it hard core, you know? Like, you don't mess with the spirits, man."

"Wow Ms. Afa, you are learning a lot about the Macushi! That is brilliant!" Noel reached over and put her hand on Afa's hand. Afa appreciated that gesture.

"Thanks for sharing that, Johnny." Afa smiled at him.

Johnny guided them through wide and narrow parts of the river and stopped the boat to talk to them about the history of ecotourism on the patch.

"In the 1990s, the forest was legally the forest of the Macushi, who had all rights over the resources on the land, also managed by the Ministry of Amerindian Affairs and the Environmental Protection Agency. The Macushi received a grant from the Commonwealth to build all three structures of the science station and ecotourism sight. The goal of it all was to keep the forest sustainable for the Macushi by sharing the forest with tourists and scientists for a cost."

"Wow, that's amazing," said Afa.

"Yeah, I wasn't alive yet, but I have heard from my parents and grandparents that it was an amazing time for the Macushi. To be able to build our own vision with funding by international foundations and non-government organizations to pay for start-up costs. In 1996, the first group of scientists stayed with us and a few months later, the tourists arrived. They were mostly British because Guyana was a part of the commonwealth, and Guyana is an Engish speaking Amazon country. Then money started pouring in, which helped the Macushi sustain their traditional way of living, while upgrading in ways they wanted to, such as getting WiFi, internet access for our guests, stuff like that."

"My, my, my Johnny, I think that is the most I have ever heard you speak! I didn't even know you knew all that!" said Noel with a loud laugh. Johnny smiled and looked down, rubbing the palms of his hands a little anxiously.

"Well, I guess when I am with the elders I hear things," concluded Johnny. But he didn't say anything else for the rest of the ride. When the ride ended, Afa and Noel thanked Johnny for an informative and eye-opening tour.

"That was great, Johnnyl! Does every ecotourist get that experience?" asked Afa.

"Yes Ms. Afa, it's included in the package."

"But not all the guides are as knowledgeable as Johnny." Noel playfully winked at Johnny, who smiled and looked down as if too shy to respond to her compliment.

Afa and Noel walked to the top floor of the round building where there were no real walls, just a wooden roof attached to a railing by narrow planks of wood. From this vantage point they had a view of the whole compound and the river gaping wide at the bank of the shore. Here, refreshments awaited. Various fruits, snacks, and beverages. There was always complimentary coffee, weak as it was, with a no-nonsense bowl of evaporated milk. Fresh milk would never survive the hot trip through the savannahs, as all supplies were brought in on that yellow van.

Although it was late in the day, they poured a cup of weak coffee while they waited for dinner.

The scientists, Afa had learned from Noel, were from all over the world. And the students who were on study abroad programs were mostly from Canada. The majority of the ecotourists were from the U.S., Europe and Asia. That Guyana was English-speaking was a real selling point for many tourists. In recent years, wealthy ecotourists had been scrambling to see what remained of the lush and wet rainforests. Such ecotourism had flooded the area with a steady stream of money and that had allowed for more resources for the Indigenous Peoples.

As they were sipping their weak coffee, Marcia walked towards them. Afa stood up to meet her, and Noel followed, standing and greeting her with a warm smile.

"Hello Marcia! Noel and Johnny just treated me to a great boat tour."

"Wonderful. That is part of the magic of the Macushi patch. I'm glad Johnny was your guide; he's our best."

"Oh, yes. His stories made the trip even better." Afa nodded.

"Truly, he is a legend. As you might be aware, we are one of the largest patches left in the Amazon, and we have an industry that is completely managed by Indigenous Peoples. Very rare, and as you can imagine, we are quite proud of that." Marcia beamed. "The science station where you're staying is my favorite building. And I'm so pleased that you're participating in our ecotourism while you are here. It gives you a chance to see first-hand how we are now self-sustaining, without any government or foreign assistance for decades." Marcia's pride in her people and this patch was evident in the way she spoke and how her eyes glistened as she referenced what they'd achieved.

"I can see in just one day that you have built something amazing here, Marcia. Something to be proud of. Noel tells me that I will have the botanical tour tomorrow, as well, and I'm really looking forward to that. I'm really curious about the plants and trees," said Afa.

"Oh, yes, on the botanical tour, you will learn about different plants and trees we have traditionally used in healing. You should also visit our clinic. It's a state of the art – like a clinic you would find in the capital city. We have a full-time doctor and nurse on staff who have been trained in Georgetown," said Marcia, beaming from ear to ear. She lit up as she described the compound, which included a children's nature club comprised of all the Macushi staff member's children and a food program that consisted of indigenous chefs who cooked the traditional meals. Her pride and esteem for the place was palpable, right up until she excused herself to speak to some other guests.

Many hours later, after Afa had taken a long nap in her room, afraid to walk by herself on the trails because of the snakes, the smell of dinner lured her, and she ventured out. As she climbed the stairs, the chefs were bringing out the hot deep silver metal pans of food. The aroma of spices she could not locate in her olfactory memory wafted up to greet her, and Afa made her way over to a group of scientists speaking with Noel. "Noel…sorry to interrupt your conversation, but what is that smell? What are they serving? It smells amazing."

"It is a local indigenous meat stew called "pepperpot." Tourists all marvel at how it is almost black in color! And we also have "cookup" which was a rice dish both influenced heavily by the African slaves that had been brought to Guyana by the British. There is also boiled cassava root and fresh and fleshy papaya to eat with dinner," said Noel.

Afa's mouth watered listening to Noel describe the meal, and she got in line, her stomach rumbling in anticipation. The chefs served the guests from trays that kept coming from the kitchen with more food than seemed possible. Afa piled her plate with a bit of everything, anxious to sample each dish, then she looked around for a place to sit. There were many tables to choose from, but Afa had noticed in the time that she had been there that the ecotourists seemed to stick together, as did the scientists and the indigenous staff. She wasn't entirely sure where to sit. Technically, she was an ecotourist for the first part of her stay, so she had been sitting with the ecotourists, even though she was sleeping at the science station with the other scientists. Hesitating, she decided to stick with what she did last night, so she walked over to a worn plastic table where the ecotourists were. They all ate dinner slowly, maybe because of the thick heavy humidity which slowed down the body, thought Afa.

After dinner, Noel asked Afa to walk to the riverbank at the edge of the clearing. There was a light rain shower. When they got to the edge, Noel turned to Afa.

"So I wanted to ask you. Do you have a feeling about what the Macushi are going to say about the rain program coming to the Guyanese Amazon?" Noel asked.

"Umm...I don't know, to be completely honest. I don't know. I'm not sure how they feel about being a part of the agreement. I want to understand their culture. I feel I can connect with them on a deeper level if I understand them better," Afa said, wondering why Noel was asking. Did Noel already know the answer? Had she talked to Marcia about the program?

"Well, I can't speak for the Macushi as I am not a Macushi community member. But personally, I think it would be crazy to turn down the agreement. It would be suicide. There is no good reason to turn it down. The patches are getting smaller. With a rain program, the patches could grow larger and expand. This is what is best for them, but also for the world. The whole world benefits from a wet rainforest! I guess it's your job to illustrate that, make sure they see the whole picture. Those swaths of dead trees that we drove through were so depressing," said Noel. Afa noticed that for the first time since she'd met her, Noel had turned serious, not as light as before.

"I know, I know, I get it, and I feel that way too. But at the same time, I feel it's important to hear them and validate their views. It is within their human rights to give or not give their consent. The world stood by as Indigenous Peoples were slaughtered centuries ago, and now suddenly the international community needs something from them, and they are expecting that Indigenous Peoples do what is in the "world's" interest," said Afa.

"The Macushi need to understand that what is best for them and best for the world are the same thing! Rain is good for everyone," said Noel with a heartfelt emphasis.

"I hear you, Noel. I'm just wondering about other factors, like culture. The Macushi spirits may stand in the way of the rain agreement," said Afa.

"Well, I'm not in a place to give you advice, nor did you ask. But if I was doing your job, I would make sure that this doesn't turn into "folklore vs. science," Noel said while forming air quotes. "That is a common barrier for us scientists. I just think it's important to show them that science IS good for them, regardless of their folklore stories," Noel said.

"Yes, I understand what you are saying. Thanks," said Afa. Noel made the situation sound so easy. Only a few generations of Indigenous Peoples legally owned their land in the Amazon, and it was only a few generations that had been able to make decisions that affected their land, or reject projects like dams, gold extraction, petroleum exploitation. So now that they had the power to decide about the rain program, of course they would be suspicious – not necessarily of her, but of the UN, the program, the entire outside community. Did any other institution or country that wanted something from them have their interest in mind? Afa had empathy for their suspicion. But it was important to convince them that without the science of the rain rocket, they would have to leave their home when the patch dried up. And if anyone understood what that was like, it was Afa.

Later that night, Afa lay in bed, practically naked, with the mosquito net lazily draped over her. It was too hot and sticky to wear anything more than her underwear. She was awake, thinking about how she would approach the Macushi. She knew what she was expected to do – get consent. But she wanted to discuss the spirits with them. She also wanted to discuss the science. She could not anticipate how it would go. She sighed heavily and flipped over onto her side. She lay there listening to the sounds of the rainforest, the light tapping sound of the rain on the thin roof, eventually lulling her to sleep.

***

The next morning Afa woke, showered with the frogs and drank sugary, weak coffee. After breakfast, she walked through the humid dense forest with the Macushi botanist while he gave a botanical tour to a group of Canadian students on the Macushi patch. They walked beyond the clearing into the thick jungle where it seemed that every step they took, there was a plant that had a medicinal quality. Afa marveled at how the local water cycle was intact in the jungle despite being surrounded by dry savannah on all sides. Noel had told her that Vishnu would arrive by the time her botanical tour finished and she was looking forward to talking to him. She felt like he understood her position more than anyone. Tomorrow was the day of the meeting with the Macushi and she hoped to discuss it with him.

When Afa got back to her room after breakfast she looked closely at her big toe and saw a black circle at the tip of the toenail. Her toe had been painful all morning, and she'd wondered when she'd stubbed it. Her simple brown leather sandals didn't offer much protection, but she couldn't remember having hit it against anything. Afa shrugged and thought she must have bruised it, and that it would go away soon. She continued with her day as planned, going on a guided walk with an indigenous guide, up a hill across the river, deep through a mangrove forest. She felt her toe every time she stepped down, especially walking downhill. A few hours into the hike, she looked down at her toe when they were crossing the river back to the station and the black spot looked a bit bigger. Hmpf. Strange. I wonder if I got a cut that's infected.

By lunchtime, the pain had increased and every time she stepped down, her entire foot throbbed. The rest of her toe had become a slight purple color. Reluctantly, she admitted it was time to go to the clinic.

The doctor himself greeted her when she walked into the small jungle clinic. He was a cheery older gentleman, who was perhaps a mix of both Macushi and English ancestry. It was clear he'd been handsome in his youth, though time had whitened his hair and slowed his movements. Still, he was charismatic. He was busily moving around his small office, which looked a lot simpler than any other clinic or doctor's office she had seen. The paint was peeling off the wall near the windows and there were old anatomy posters in faded pastel colors on the walls. "How can I help you, young lady?" the doctor asked.

"Hello, Doctor. I am staying at the science station, and I wanted to show you my toe. I've got this black circle here," Afa pointed. "And it hurts." She raised her foot, took off her sandal and plopped her foot on his desk. Her toe looked like a purple light bulb with a black spot at the top of the bulb. "It feels like needles and pressure." The doctor only looked at her foot for a few seconds before he put his hands on his waist and looked at her with sober and knowing eyes.

"Ah, yes, you have a parasite in your toe that we call tunga! It has crawled into your toe and laid eggs. You must have gone to the dry sandy part of the river. I will have to remove the eggs from your foot." Despite the doctor's matter of fact delivery of this news – as if he saw eggs in people's feet every day – Afa's jaw dropped and her stomach turned upside down.

"A parasite? In my foot? Can it go anywhere else or is it just there? And how will you remove it? Will it hurt?" asked Afa.

"No, it's just in your toe. It won't hurt – at least not while I am doing the procedure. I will use dry ice to freeze and numb the toe. Then I will scoop out the eggs with an instrument. Then we will bandage the toe. You should wear closed toe shoes while your toe heals. It happens all the time here. It happens to the boys that drive the boat tours as they are always in the sand. I can do the procedure right now if you like?" said the doctor.

"Yes, please, do it now!" said Afa, horrified at the idea of anything hatching inside her body if she waited longer.

As he had described, her toe was frozen with dry ice, mercifully, so she didn't feel him scoop out a small part of it. But she couldn't bear to look when he was finished. After he'd bandaged her up, she thanked him and hobbled out. It felt fine to her until her toe was no longer frozen, then it throbbed painfully. Where he had scooped out her toe, it ached sharply. She limped back towards her room with a slight hobble.

Afa was just inside the science station when she spotted Noel.

"Ms. Afa! Why are you limping? The bandage on your toe... oh, no..." Noel covered her mouth with a deep look of concern and shook her head slowly.

"A parasite laid eggs in my toe and the doctor scooped it out. It hurts."

Noel nodded sagely. "You have had tunga! You are now a local." Then Noel shrugged it off and claimed matter of factly, "But don't worry, Ms. Afa. You will heal in no time!"

"Thank you, Noel. I feel kind of grossed out."

"I understand. It happens to the best of us. But hey, your friend Vishnu has arrived. I am just walking over to greet him. Do you feel well enough to come with me?"

"Yeah, I think I can manage." Afa didn't want to seem too excited, like a teenager with a crush, which was how she felt. In fact, all she really wanted to do was lie down. But Vishnu's arrival had lifted her spirits somewhat – she even thought her toe was throbbing a bit less. A little endorphin rush, perhaps, to elevate her pain tolerance. She wasn't going to lie – Vishnu had lit up all the requisite parts of her brain in order for her to find someone attractive. Her crush on Vishnu was all about his mind. She had loved listening to him talk about the Macushi. Of course, it didn't hurt that he was tall and good looking, either.

"Vishnu! How are you? Nice to have you back. We missed you. You have met Afa, of course, our friend from the United Nations," said Noel.

"Yes. I have had the pleasure of meeting Afa," said Vishnu with the swagger, style and grace of a man who was very comfortable in his own skin. He was looking at Afa while he responded to Noel. And his gaze remained on her as he said, "Ms. Afa, I've been thinking about you."

Afa felt her cheeks grow warm. "Oh, really? Why?"

"Because you make me want to break out my guitar. Do you know the song Black Magic Woman by Santana?"

"Of course." Afa laughed, a little confused, but certainly intrigued.

"You make me think of that song. Yes, absolutely. You come here promising rain, this beautiful, mysterious and powerful and dangerous woman. You are a black magic woman!" Vishnu hummed a few bars of the song, and Noel held up her hands, swaying and grooving as he did, a big smile on her face.

"Yes, yes, I can see what you are saying, Vishnu. She is!" Noel nodded in agreement.

Afa laughed at the pair of them, shaking her head. "Thank you, but I don't know how dangerous I am…and I don't feel too magical with my bandaged egg toe." Afa was making light, but in truth, her belly burned with excitement at Vishnu's assessment of her. She didn't understand quite what he meant, but it thrilled her to know he had been thinking of her. And even more thrilling was that he'd assigned such a sexy song to her.

"Oh, I think you are dangerous, Ms. Afa," Vishnu said, leveling her with an intense gaze. His eyes burned into her, and she didn't know what to do with herself, where to put her hands. She just fumbled about, so obviously not suave or sexy with her big, bandaged toe throbbing, and standing next to Noel who exuded raw sex appeal. Yet here was this enigmatic man treating her like she was the only woman in the room. Perhaps the connection – and attraction – she had felt on their initial drive in together hadn't been one-sided.

"Oh, leave our Ms. Afa alone, Vishnu!" said Noel with a knowing smile. "For sure, we are all going to have fun tonight! Our supplies are here so we have rum to drink, and we are going to play music for all the guests this evening." Noel's spirits were high, and she clapped her hands to show her excitement.

"Vishnu! You have come to see us again," Marcia called, coming around the corner from the science station. "We are so happy to have you!" Afa had never seen Marcia this expressive before. It was clear that Vishnu was a favorite of everyone here.

"Vishnu is one of us, so you can go on walks with him all around the grounds of the compound with no worries. He knows all the snakes, and he is a dear friend of the Macushi," said Noel to Afa.

"Marcia, my meeting with you and the Macushi community elders tomorrow is after lunch, correct?" asked Afa, nervous to even bring up work when there was so much enjoyment in the air.

"Yes, tomorrow. But no business tonight! We will be having a party with the ecotourists! A lot of the Macushi will be there too, and the scientists. We will be serving rum mixed with coconut water, a local favorite. And Vishnu has brought the music!" said Marcia.

"I did," said Vishnu matter of factly.

"Vishnu is getting his PhD in Jamaica, so he brings us all the good Caribbean music. We have the speakers set up for tonight. We will have a dance party in the open-air top floor, what we refer to as the reception area. I know it's a bit early in the day, but we'll be serving drinks soon," said Noel, enthusiastic about getting the party started. "We'll turn it into a roof-top bar!"

"I will see you soon, Afa?" asked Vishnu, looking deeply focused on her eyes.

"Yes, yes....I..I will be there," said Afa, stammering a bit as she spoke. "I don't know if I'm in dancing form, but maybe some rum will help..."

"What did you do, by the way?" Vishnu asked, referencing her toe.

"The tunga!" Noel chimed in.

Vishnu shook his head and laughed. "I've been there. I should have known. Happens to the best of us. You'll be fine in a day or two, but it is nasty stuff. Stay off it for the next hour or two, take some ibuprofen. I bet you're right about the rum." He winked at her.

Afa smiled at Vishnu, then hobbled her way back to the science station, into her familiar concrete room with the mosquito nets hanging from the ceiling. She limped to the bathroom and passed the shower, stopping to see if the frogs were still there. They were, just looking at her as if no time had passed. Afa laughed to herself, then sat on the bed, trying to make sense of everything. She felt strangely drawn to Vishnu for his swagger as well as his intellect. He was so knowledgeable, and she couldn't deny he was sexy as hell. He had a liberated yet commanding persona. He was both respectable and fun, a hard combo to find. Perhaps that is what drew others to him as well. Vishnu had many people that seemed to love him at the compound.

She couldn't help but be excited about the prospect of seeing him tonight, getting to know him better. Whether or not she'd have time to speak with him much about business before her meeting with the Macushi, she didn't know. But tonight promised to be fun at least.

After resting with her foot up, and even a short cat nap, she changed into something more appropriate for a party. She had brought only a few clothes so the simple, cream-colored linen sheath would have to do. Her toe stood out, bandaged as it was, but the afternoon of rest and ibuprofen had done her some good. The throbbing had eased, and she was able to bear more weight on her foot now. Her hair was down, and she tousled it a bit, then wiped the beads of sweat from her forehead before she left her room for the reception hall.

The round shaped reception area was open to the air, surrounded by wooden rails on all sides. The sun was setting and there was a beautiful, peach streaked sky above the view of the moving river beyond the clearing. Rain every hour or so for the past few days meant that the air was balmy and humid. Noel greeted Afa and took her hand, pulling her to the makeshift bar. Noel was reliably so full of life – spirited, radiant and very supple. Her feminine figure looked stunning as ever in a form fitting dress, her shiny black hair pulled into a bun. Her laugh was deep and genuine, and Afa felt like she was reconnecting with an old friend as they stood at the bar. Noel spoke about her husband.

"He has his own ecotourism company that takes people to see the tabletop mountains in Guyana," Noel said.

"That's so cool! Do you ever join his tours?" asked Afa.

"No, not really. I have a very smart daughter from my first husband, and I have to maintain a routine for her since she goes to school. She is with her dad in Georgetown while I am here in the jungle."

Afa felt that even though Noel was older than her, she had a younger spirit. But Afa believed they had one important thing in common; they were both professional women. They had never let a man come before their careers. And despite Noel having an impressive education and position, that didn't stop her from having a raucous good time with the people at the compound. She had a drink in her hand already.

"One rum drink for Afa!" yelled Noel to the staff serving the drinks, trying to be audible above the sound of reggae that had just started coming out of the speakers. "Maybe you can come and stay at my house in Georgetown when you are finished with the Macushi here?" said Noel loudly into her ear. "I'm going back tomorrow night, after your meeting." Afa liked that suggestion and nodded vigorously.

"Yes, I would like that very much! Thank you!" said Afa loudly.

Noel handed Afa her drink and said, "This is very old rum, it has been aged so it tastes smooth – you will like it!" Afa had taken her first sip and was delighted – refreshing coconut water infused with a rum that tasted of burnt wood and brown sugar. It warmed her insides, and the pain in her toe ebbed a bit more, almost disappearing. She thought she might even be able to dance, given enough of this rum. Smooth, indeed.

While Noel was talking, Afa looked around and noticed that the scientists, ecotourists and the indigenous staff were all dancing. She spotted Vishnu who was also dancing, moving from one dance partner to another. Noel took Afa's hand and pulled her to the center of the room, under the dusky sky. Vishnu noticed and approached, gesturing to Noel first. He put his hands on her waist, and they moved to the beat of the music. It was humid and the rum made Afa's body fluid and sultry. Vishnu and Noel made a striking pair, dancing right in front of her, and she tried hard not to stare. Vishnu made eye contact with Afa, then pulled her in to dance with them, putting his arm around both Noel and Afa, pulling them both close, wrapping his arms more tightly around her. Noel made a motion with her hand that she was going to get another drink and broke up their little triangle.

Vishnu put his hand around Afa's waist and they continued to move together. The music ran throughout Afa's body like a deep grounding beat and she was viscerally aware of Vishnu's hands, her body catching fire where he was touching her. Her heart started to beat a little faster. She wanted him to pull her in closer, as she had watched him do with the other women he had danced with, but Vishnu kept a space between them, which created an electric current, drawing her in. He began to drop his body closer to the floor with his legs and he made a motion to Afa to join him. She felt his hand tighten on her back and a jolt of energy rushed up her spine. The reggae beat was loud, and she was moving her hips as he moved slightly closer to her. Then, suddenly, he pulled away and approached another indigenous staff member, dancing away from Afa, and taking all the heat with him.

Afa caught her breath, surprised at Vishnu's sudden departure. She flushed with embarrassment, though it would be difficult to tell, as her face was likely red from the rum, humidity and movement anyway. Her time dancing with Vishnu was over, but a lingering sense of desire filled her body. She barely knew him, but he felt familiar, and now he was dancing with all the other women. Maybe everyone felt comfortable with him, she could see why. But she wished he had danced with her for longer. His smile was infectious, and he had a warmth and charm that made him magnetic. Clearly, no one was immune to his charm, she thought, watching him now as he effortlessly moved around the dance floor. Her body hot and sticky with sweat, she went back to the bar to get another rum drink. Noel approached Afa when she was at the bar and said loudly over the music.

"Isn't this great? Everyone together like this? We have these parties every now and then, but especially when Vishnu brings the music. It's great that you happen to be here for one of our parties."

Afa scanned the dancefloor and bar for Marcia.

"Is Marcia here?" asked Afa.

"No, Marcia doesn't dance anymore. She did when she was younger, but she's probably with the elders who also don't come to parties anymore. Are you looking forward to meeting with her and the elders tomorrow? They are a special bunch of people," Noel commented.

"Yes," said Afa, sobering. She wondered if it was appropriate that she was dancing and drinking before such an important meeting.

"Well tonight, though, no business. Sorry to remind you of business. Tonight, is about the party!" said Noel gleefully, validating Afa's presence at the party. Evidently Noel saw no conflict of interest, so Afa settled in and let herself fully relax. As Afa drank her second drink, she couldn't help but locate and follow Vishnu out of the corner of her eye. She watched him dance with people he knew, people who were so happy to see him. Occasionally, he would stop dancing to talk to some of the scientists who were mostly standing around and bobbing their heads.

Afa observed the crowd of people, all from different backgrounds, dancing together in the rainforest. Almost everyone was dancing, and Afa observed that the Macushi were not as loose as the ecotourists, who were mostly Canadian. They were moving on the dance floor like wet noodles. Reggae boomed out of two large speakers in the darkened rainforest night. Noel kept Afa on the dance floor for a few hours; they'd dance a while, take breaks and talk, then dance again. Afa was amazed that her toe had ceased to throb. The rum had some powerful anesthetizing properties!

At one point in the night, Afa and Noel were dancing to a slow reggae beat, when Afa looked up to see that Vishnu was watching her. He didn't look away but held her gaze and smiled as he walked straight towards her.

"Shall we take a walk through the forest?" Vishnu had leaned in, so only she could hear him. Thrilled at his sudden attention, Afa nodded. "Meet me in front of the science station in

15 minutes," said Vishnu. Excited and nervous at the prospect of being alone with him, Afa noticed how tipsy she was. Her arms and hands fumbled awkwardly, like her body was no longer attached to her brain. As she wobbled back to the science station to change her shoes, she could hear Vishnu laughing.

Fifteen minutes later, as the sun was starting to set, she stood waiting for him at the edge of the clearing with closed toed shoes to avoid more tunga. Her toe was slightly tender, with the bandage pressed against her shoe, but if she went slowly, it wasn't bad. While she waited for him, she savored the beauty before her, breathing in the clean, humid Amazon air. When Vishnu gently touched her arm, she startled slightly.

"I didn't hear you coming. You're so smooth," Afa laughed. Was she flirting with him?

Vishnu smiled. "Or you were lost in thought. Come. Let's walk," he said. They walked in silence across the clearing, and Afa was aware that everyone could see them walk together into the forest. In the back of her mind, she worried a bit about the perception of the others. If they saw her leaving to walk with Vishnu, would they assume they were romantically involved? Afa wanted to be taken seriously. But then again, she told herself, Vishnu was friends with everyone, warm with everyone, danced with everyone.

"Are we going to walk far?" asked Afa.

"No, it gets dark in about twenty minutes, then we will head back," said Vishnu, and then he paused. "Is your toe okay?"

"Oh, yeah it's okay. You were right; ibuprofen and rum and I can barely feel the pain now."

Vishnu laughed and nodded, "Not much that Guyanese aged rum won't cure, that's for sure."

Afa was pleased that she had 20 minutes alone with him, his charm all to herself, that she didn't have to share his attention. His body was graceful, and he moved easily. She watched his calf muscles on his brown legs as he walked ahead of her. He hung back so they were walking side by side and he focused on her with his eyes, his smile broadening as he got close to her. Afa walked with him across the clearing and onto one of the many paths through the wet forest, her feet flattening the wet plants of the forest understory.

Suddenly Vishnu stopped and said, "Look! Look! Do you see it? Hold on, I will give you a better view." He then bent down, reached into some plants, and picked up a snake. He held it right in front of Afa's face. "Don't worry, it's non-venomous. It's harmless."

Afa gasped and drew back, frightened.

"Oh!" She exclaimed. "That's…that's okay. I don't need to see it closer." She kept backing away, widening the distance. She trusted that Vishnu knew a harmless snake from a deadly one because he was a biologist and had a deep understanding of the forest ecosystem, but she still didn't need to get any closer to the thing.

"Ok, can you put it down now. I think it's looking at me. They can probably smell fear."

Vishnu laughed. "It's ok. I wouldn't pick up a snake I didn't recognize." He put down the snake and they continued walking. Afa let out a deep sigh of relief.

"Are there any plants around us that are used by the Macushi that you can tell me about?"

"Well, yes. Let me see, here is one. See this grass here? It is called Parawian. Macushi believe that it allows people to run fast and far. They believe that they can change their own body's abilities with plants. To use the Parawian, you must collect and burn the grass and apply the ashes to cuts made at the ankles and at the side of the knee joints."

"Does it work?" asked Afa.

"I don't test the medicines – that's not my job. I just make a record of all the plants that the Macushi use to transform their bodies as they see it. When the elders take the ecotourists on the tours, they show them their different medicinal plants, but they ask the ecotourists not to harvest anything. My job is to document their perspective of the plants on the patch."

As Vishnu spoke about matters of the Macushi, so careful to preserve their perspective, Afa's attraction grew. The admiration she had for him seemed to flow through her body from her feet to her fingers to her mind; something beyond physical. She thought about what it would be like to take his hand, but obviously, she resisted. Theirs was a professional relationship, despite the rum and the flirting and the dancing.

"I can keep coming back here. Hell, I could live here, but I have my schooling in Jamaica and my family in Georgetown. Maybe one day I can get a permanent post working on the compound."

"Seriously. I would love to live and work here," said Afa, looking at Vishnu with bright eyes.

"Yes! You would be great! What if you and I both worked at the compound? I could work with the biologists, and you could work with the relations between the Macushi and the international community. We would be a great team," his eyes were wide as he looked to Afa for a response.

"That's a nice fantasy, but I could never. That type of position should always be a Macushi community member."

"Yes, of course," said Vishnu, returning to a more tempered expression. "But, fun to think about, no?"

Afa grimaced; she'd dampened his enthusiasm. But she thought Vishnu's suggestion was wild. It was fun to imagine a dream job where she could leave New York winters and live here in the forest, working in some capacity within the science station. But she wasn't a scientist or Macushi, so what would she do? Her pragmatism wouldn't allow her to run away with a wild, fantastic idea. Vishnu was passionate and said what he was thinking, no matter how unrealistic and improbable it was.

"Speaking of the Macushi, I'm having a hard time planning for my meeting tomorrow, I don't know how to approach it," Afa admitted, trying to recapture some of the ease between them.

"Ah, well, to be honest, I do think it will be a difficult meeting. I have heard in casual conversation that there are quite a few community members that don't want the program here. The shaman and Marcia have a lot of influence over the community, so I don't know what will happen, I'm not sure what they are thinking. And I feel for you. I can see your struggle. I can see how hard it is for you to reconcile what you are here to do for the UN with your respect for the Macushi worldview. I know your colleagues at the United Nations will be disappointed if they don't sign the agreement," said Vishnu

"You understand, then. It's going to be very hard to go back to the UN and tell them that I failed to get consent, if that's what happens. Although everyone at the United Nations respects their rights, naturally they all hope that they would agree to an Amazon-wide program since this is all one interconnected forest ecosystem. They're counting on it. The stakes are high," emphasized Afa.

"A lot of international scientists want to preserve the Amazon forests. But it's not just important to preserve their forest, it's also important to preserve the worldview of people that live in the forest, and to respect that view, no matter what it is. Like we discussed in the van on the way to Georgetown. No matter how unpopular or controversial it is among global environmentalists, it must be respected," said Vishnu.

"Exactly. My boss is a scientist, and I don't think he will understand if their decision is to not sign for cultural reasons."

"Listen Afa, I think you are doing a great job. I know there is resistance, and I don't know how much resistance – but I know they believe you are doing your job and have good intentions."

"You think so?"

"Yes, definitely. I work with them. I can tell."

"I don't know. I struggle with this," said Afa. "I don't know if I am doing good work. I feel like I don't want to apply pressure, but my boss says that's part of my job. But how do you pressure people to go against their beliefs?" The words tumbled out of her, a freefall of emotion and vulnerability.

"Afa, you are doing good things here. What if, by honoring the Macushi worldview, that is succeeding, regardless of whether they sign?"

Afa smiled at Vishnu, his words lifting some of her angst. "Thank you, Vishnu. Your reassurance means a lot to me." They walked in silence, the only sound the leaves that brushed their arms and legs, the earth that yielded beneath their feet. All the lush surroundings were moist, and the plant material was wet on the earth. They continued to talk and walk while the sun set, their eyes becoming accustomed to the dark. As they walked past the bungalows set up for ecotourists, which Afa had noticed on her first visit, she wondered about them; what were they like inside and how must it be to stay in one?

"Are these bungalows expensive to stay in?" asked Afa as they walked by.

"Yes, they are. That's why it's always wealthy Europeans and Americans staying in them," said Vishnu.

Vishnu stopped walking and pointed at one eco-tourist bungalow with the door slightly ajar. The keys were in the door and the windows were open. He took Afa's hand and started walking towards the bungalow in the dark.

"What are we..." Afa started to say.

"This bungalow is unoccupied, it's empty. No tourists. Come on, let's get into the hammock on the porch!" His smile mischievous, his voice full of glee.

"I...I... don't know if this is right, Vishnu..." Afa looked around, expecting a maid – or anyone really – to catch them.

"Come on! Let's experience the luxury lifestyle of an eco-tourist!" laughed Vishnu, his voice taking on a hushed whisper. He climbed into the hammock and held it open for her to climb in. "Come on, get in!" he said, still whispering.

Giggling, Afa gave in. It was as if she was compelled to join him to do something risky, like some force of attraction was urging her forward into the hammock. She gingerly climbed inside. Their bodies rolled together like a California burrito as their weight brought the sides of the hammock up and around them. Afa tried to adjust, but the movement just smashed her into Vishnu even more, so she contented herself with lying basically on top of him, swinging in the cotton hammock suspended from wooden beams, overlooking the river and under the canopy of trees. All around them the cooing and cawing of birds and monkeys echoed in the cracking jungle branches. Afa sighed and thought about how much she admired Vishnu's passion, that she loved how deeply he understood the indigenous perspective.

After a few minutes of lying still, touching him with one side of her body, she felt his hand rest on her arm. Her eyes had become accustomed to the dark night sky and she turned to face him. He met her eyes with his own. Their faces were close, and she was overcome with an intense feeling of wanting to be even closer to him. The heat of his body, the feeling of his breath tangling with her own – it was a heady, heady feeling. Afa's heart quickened when Vishnu leaned in, ever so slightly.

"Do you want me to kiss you?" asked Vishnu, his voice a whisper. Yes, thought Afa, but at the same time, a thousand thoughts rushed into her head about why she shouldn't. Blood rushed into her cheeks, flushing her face with heat and color. She could feel his attraction to her mind, not just her body. But as colleagues, would a fleeting act of passion make things awkward? Overcome with logic overriding the heat in her body, with some effort, she turned her face away from Vishnu. She really, really did want him to kiss her. But she couldn't let it go there. Not now.

"No. I mean, yes," she panted. "But I really admire you, and I want to work with you, so... we can't."

Vishnu drew a deep breath, then leveled her with an intense gaze. He nodded. "I get it." His tone held defeat and disappointment.

They lay together like that about another minute, a bit awkwardly, when they heard footsteps in the forest.

"We have to go!" said Vishnu in an urgent whisper. With some struggle, he rolled out of the hammock and took Afa's hand to help her out too. Adrenaline flowed through her body. What if they got caught? They hadn't even done anything, but getting caught this way wouldn't look good. Colleagues might think... well, she didn't know what they would think, but she needed to keep things at a certain level of professionalism, especially considering she had a job to do here. They walked furtively back to the clearing in silence, their bodies close together as if they wanted to touch but resisted the temptation.

When they got back to the science station, they said goodnight and waved goodbye in plain view of the other scientists, Afa attempting to reset the charge between them. Once she was back in her room, she hurriedly changed out of her clothes, brushed her teeth and got in bed. She felt like she was in a daze, moving through the motions, reliving the rush from what took place in the hammock, imagining what it would have been like if they had kissed. Lying on her bed in her room, she could hear Vishnu talking to other scientists outside the building. She was so sleepy from the rum, she began to drift off to sleep with Vishnu just a few feet away from her door.

"Afa, Afa – come with me, I want to show you something," said Vishnu.

"Ok, what do you want to show me?" said Afa.

"You will see, follow me."

They were walking across the clearing. It was late at night, eerily dark and silent. There were no other people anywhere to be found. She silently followed Vishnu, and they were moving so quickly, as if floating.

"Where are we going?" asked Afa,

"To the river," said Vishnu.

In no time at all, as if transported, they were at the river's edge. Afa was peering into the water, barely able to see anything underneath the black sky. Suddenly Afa saw a woman emerge from the river, coming to the surface of the water, resting against the riverbank. Her face looked indigenous, with almond-sliver eyes, a narrow chin and high cheekbones. Her hair was straight and black. When Afa looked closer, she could see that beneath her breasts, there were scales on her body, like a fish.

"Who are you?" asked Afa, puzzled over what she was seeing.

"You know who I am. Come a little closer. I want to tell you something," said the woman with the scales, her head bobbing up and down in the river.

"I don't want to come any closer!" said Afa, suddenly feeling frightened.

"You must come closer; I have to tell you something important."

Despite Afa's fear, she felt a pull towards the riverbank, like a magnet. She gingerly stepped to the edge of the river, her toes almost touching the water. With no warning, the woman grabbed Afa's ankles.

"What are you doing? Let go of me!" Afa was filled with terror.

"You must come with me," urged the woman in the water, her voice taking on a soothing tone even as she pulled harder.

"Let go! I don't want to come with you! Let go of me!" Where was Vishnu? Suddenly, Vishnu was nowhere. She had no one to help her. She felt the strength of the woman pulling her in at the same time as she was losing her grip, her fingers slipping on the banks, absolute terror and fear making her whole body shake.

"It's ok, follow me," said the woman soothingly. Then, becoming sharper with Afa, she said, "You must come!"

"No, no! I don't want to go with you! Stop! Stop!" Afa felt the water at her feet. Let me go!" She reached her hand up towards the bank and slammed it down hard, trying with all her might to regain her grip.

Afa sat bolt upright in her bed, sweat covering her whole body. The sheets were damp, and the mosquito net was tangled at her feet. Her heart was a drum in her chest, her breath caught in her throat. It felt like a panic attack. She gripped the edge of the bed, reminding herself she was, in fact, in bed and took deep breaths to calm down. She reached for her water bottle on the floor and guzzled. She tried to comfort herself by muttering under her breath, It wasn't real. Just a dream. It's over. Her heart rate resumed its normal beat, her breath calmed, and eventually, the consistent, reliable sounds of the jungle lulled her back to sleep.

The next morning, Afa woke up feeling uneasy – her stomach knotted up with nerves. The dream had stayed with her, and she knew she did not want to apply pressure to the Indigenous Peoples to abandon their Water Mamas and join the rain program. She wanted to respect the Macushi worldview, as Vishnu had called it. She respected their belief in Water Mamas, and she wanted them to benefit from the rain program. She just wasn't sure how to accomplish both.

She got dressed and walked over to the round building for breakfast of papaya, bread, eggs, and weak coffee with evaporated milk and sugar. She glanced over to see Vishnu coming up the stairs and heading for the line for breakfast. After their heated dance and their time in the hammock, Afa wanted to be near him. She realized that since she met him, she'd always been aware of his presence when they were in a shared space. Afa saw Noel was already seated with the scientists, so she went to join her.

"Good morning! You were really dancing last night! You are a good dancer, girl!" Noel shimmied her shoulders at Afa.

"Thank you! You are too." Afa laughed. "We move our hips a lot in South Pacific Island dancing. I grew up dancing like that," said Afa. Peripherally, she was aware that Vishnu was seated near them but at the other end of the table, and her heart rate picked up when she realized he had started to gather his things, moving closer to them. Suddenly she didn't know how to act. She was aware that words were still coming out of her mouth as she spoke to Noel, but also that she had no idea what she was saying. Had Noel asked her a direct question? Was she sweating? Vishnu was having a physical effect on her. Finally, she made eye contact with him when he sat down, across the table from her.

"Good morning, Noel, Afa." Vishnu nodded at them. "Today is your meeting with the Macushi. How are you feeling about it?" asked Vishnu.

Considering that I could barely make conversation with Noel just now, not great, thought Afa. "I'm nervous," she admitted. In her opinion, being nervous was a sign of incompetence, and it pained her to appear incompetent to Vishnu and Noel. But at this point, Afa was starting to question her inner compass and she felt it was ok to be vulnerable with her new friends.

"Yes, the stakes are high!" said Noel with a dramatic emphasis. She turned to Vishnu and said, "If she can't negotiate a contiguous Amazon wide rain program, we..."

"Yes, I understand the stakes, Noel, I am a biologist too," said Vishnu, cutting Noel off. Then he looked past both women into the distance and seemed to disengage. Afa registered this, but went on.

"I know the stakes. We need rain here. We need rain everywhere. Rain agreements have been negotiated in the other two contiguous rainforests in Asia and Africa, but the Amazon is the biggest and has the most influence over global climate. I get that. I really do. I really need consent." She looked at Vishnu, but he was still distant.

"My advice is to appeal to the elders," Noel said. "They hold all the power. I used to get intimidated by a room of Macushi elders when I first started working at the compound, but everyone is so kind here. No matter what happens, they won't be disagreeable with you, Afa. Plus, this is your second meeting with them, there will be familiar faces, and Marcia will be there! Maybe Marcia will respond to you making a case for a wet Amazon helping to restore rain to the whole planet. I always saw her as a global thinker. Maybe you can count on her – I remember she wasn't at the last meeting," said Noel.

"Yes, Marcia will be there," said Afa as she stared at the fleshy orange papaya on her plate. Vishnu, who was a little hard to read this morning, changed the subject and started to discuss the visit by a group of Canadian scientists. The conversation turned to logistics at the science station. Afa tuned them out and ate in silence. Why had Vishnu abruptly changed the topic? She tried to manifest and visualize herself confident in the meeting, getting consent. All those self-help books she had read on manifesting, she was mentally drawing from those books at this moment. Vishnu paused in his conversation with Noel and was staring at Afa.

"You are awfully quiet, Afa," said Vishnu.

"You guys are discussing business…" Afa shrugged. "So I thought I would let you work it out."

"That's true. Please excuse us for talking about work and not including you," said Noel.

"It's ok! No problem, I have to go back to my room and get my laptop for the meeting so I'm going anyway," said Afa.

"Hey, Afa, let's talk about the meeting after you are finished. This evening before you leave. Noel said you guys are leaving as soon as it gets dark. I will try to find you after your meeting."

"Sounds like a plan. Thanks Vishnu." Afa was happy about the prospect of being able to process with Vishnu. He would understand the outcome, no matter what.

***

As scheduled, after lunch, Afa walked into the large meeting room to meet with about twenty Macushi community leaders and members. Her toe had a fresh bandage, and though it was still a bit painful, she could at least walk without hobbling.

"Ah, great! You are here. We heard you got the tunga! You are one of us now!" said Marcia with a look of empathy that turned into a knowing smile.

Afa nodded and looked down at her foot. "So everyone keeps telling me."

Marcia turned to address everyone else and said, "Let's begin."

The Macushi elders were already seated, and Afa glanced around the circle. She was surprised to see some of the younger members of the tribe there too. Noel, who was not Macushi and lived in Georgetown part time, had gone home after breakfast and wasn't present. Afa would miss her friendly, encouraging face as she presented. There were also no scientists in the room, just the Macushi. A few of the Macushi ecotourism staff were present as well. Afa inhaled a long, deep breath and mentally readied herself to address what appeared to her to be an all-knowing room. She took her place at the center of the circle as the room quieted.

"First of all, thanks for hosting me the last few days. The ecotourism has been lovely, and I have immensely enjoyed what your patch has to offer. As you know, this is my second meeting with the community, but I see some new faces. I thought we could do introductions again?" Afa raised her eyebrows and looked at Marcia for approval.

"Yes, I think that's a great idea," said Marcia, who directed the attendees to say a little about who they were, what they did at the compound or jobs they had on the patch, and give some context about their relationship to others.

One by one, the Macushi people spoke, and it proved to be a room full of diversity in both their backgrounds and roles they served. Two younger, skinny men in faded and worn cotton t-shirts and plastic sandals led the boat tours. The botanist and healer, whose face was gentle and kind, but weathered, was one she recognized. His wrinkles had hardened around his brow and in between his eyes. She also recognized a young Macushi teenage girl with jet black hair and heavy makeup who worked in the kitchen doing meal prep. And there was an older woman with gray hair pulled tightly into a sleek knot she had seen cleaning the rooms around the compound. And of course, there was the doctor who had helped Afa with the tunga in her foot. All were present, and it gave Afa a feeling of being among familiar faces, perhaps friends, but at least community. Her nerves settled just a bit.

"Why don't I start by very briefly reviewing what the program would do and what signing the agreement means for your community. We went over most of it in our last meeting when I was here a few months ago." Afa took just a few minutes to re-explain the framework and details of the multi-country program, and what restoring rain would mean for the patch. "As you are aware, at one time, the Amazon spanned seven contiguous states, and the UN is trying to facilitate a rain program that would cover the entire ecosystem. We can only treat the Amazon as one whole ecosystem if all the tribes consent to the agreement. If a tribe does not consent, it will be hard to avoid certain patches, and the rain program might not work to restore the entire ecology of the rainforest."

Afa also explained the importance of Free Prior and Informed Consent that was awarded to them by the UN. "Of course, it is within your human rights to refuse the rain cloud seeding program since this is your land." She paused. "Now that you know the benefit of the program, you also need to know what your rights are. So please, let me know if you have any questions."

A bead of sweat trickled down her back, and her unease over trying to successfully present to this community both sides of such an important situation caused her stomach to twist. There was so much riding on this, she wanted to be sure she had disseminated all the information. She hoped that was the right approach. But it was done now. All she could do was hope she'd done her job well.

The doctor who had fixed her toe raised his hand and Afa nodded in his direction. "Let's discuss the chemicals that will be used to make rain."

"Yes, absolutely. Your perspective on the chemicals is important, let's discuss." Afa's hands had curled into tight fists, and she took a deep breath, attempting to dislodge the brick that had suddenly settled onto her chest. This was the big question, and she knew it was coming—of course she did. But it was important that she reframe her reaction, that she stay calm and answer his questions from a place of informed, educated, and reasoned perspective. Inside, she tried to compose herself as she thought, why am I so nervous?

"Afa, we benefit not just from managing and owning our own land here on the patch, we also benefit from the science that the scientists bring and share with us. We have talked to scientist visitors who had told us about the rain program ahead of time, before you came to visit with us. They spoke to us in more detail about the chemicals that come out of the rockets.

We knew before we met you that silver iodide is shot out of the rockets into the sky. We also know that silver iodide can cause illness, fatigue, injury in people, and it also enters our water. We are concerned about our land if it is exposed to this chemical." The doctor's words were firm, and she immediately felt the climate in the room shift to one of solidarity with the doctor. People were nodding their heads and wrinkling their foreheads in concern as they listened to him speak.

"Ah, yes, you're correct, and I'm happy that you've done your background research." Afa began. "But let me assure you as well that it has been proven by many scientists that in small amounts, the exposure is negligible, and the chemical does not cause harm. The majority of global scientists believe that the amount of silver iodide used to make rain is safe. I assure you that..."

"Ms. Afa, we are aware of the Australia case." Marcia was now speaking up, interrupting Afa. Her tone was stern, as if irritated by the direction Afa had been headed. "We know that in Australia the claims of negative environmental impact were disputed by the International Weather Modification Association. But we know that there are scientists in Australia who are still working to prove that the silver iodide is harmful to water, vegetation and people." Marcia concluded. Afa was stunned. Marcia had been so friendly and kind towards her, and this felt like such a hard departure. Internally, Afa was impressed by Marcia's knowledge of the Australia case, which was always a thorn in the side of any scientist promoting the rain program. The doctor again raised his hand to speak.

"If I may, Ms. Afa, we know that in the U.S., silver iodide has a hazard rating of two. It is not a rating of zero, is it? Two is not the same as zero. We here feel that we want to take zero chances."

"I understand your perspective. However, are you not concerned about the reality that your patch may become smaller? The risks that come with the shrinking of the patches?" asked Afa. "No rain is also a risk of zero chance that the forest will come back."

"Yes, we understand that risk as well, and we think," he looked around at his community. "We are willing to take it."

Afa felt the blood drain from her face. She was nodding, wanting to validate this reality, but she was devastated. "Yes, of course, I understand," said Afa, her voice small. She was overwhelmed with emotions. Hearing them speak about the patch had made her think of her island that was sinking into the sea and how her parents fought about when to leave. Her father had ignored her mother's instincts, believing instead that the sea level would stabilize. But her father had been wrong. Sea level kept rising. The Macushi were taking a risk, the risk that their patch would get smaller because they would not take part in the rain program, because they feared being poisoned. The emotions of rejection and failure and incompetence and betrayal competed for space inside her, and she fought to keep her face from showing every single one.

Afa was struggling to find a balance between hearing the community out, supporting them, while also pushing her UN agenda more firmly than she had before. Personally, she wanted to respect their autonomy and self-determination as a tribal nation. Professionally, she was there on behalf of the global environmentalist community to try to obtain consent amongst the tribes. As her boss kept reminding her, the choices of the Macushi were not just affecting them, but rather the entire atmosphere. The world's population benefited from the expansion of the largest wet rainforest in the world. She was beginning to prepare herself to walk away without their consent, despite her belief that the program was not just good for them, but key to their continued prosperity; indeed, for the continued prosperity of the world. But she would try to explain one more time.

"Please, if you will indulge me, let me illustrate how the silver iodide would create no harm. I know you have heard differently from the scientists who have stayed with you, but most scientists around the world have confirmed that the amount of the chemicals produced by the rain program were in such small amounts, they could barely be traced. I have been to presentations of the US Health Services and Environmental Protection Agency who have conducted studies on the environmental hazards related to silver from rain programs. One of the scientists from the study told us that you are exposed to more silver in a tooth filling than you are by a lifetime of rain programs. And I can assure you that..."

"Afa, pardon the interruption. We have not heard anything new today that has affected the decision we have already made. However, before we give our final answer, I would like to say a few words that I have prepared," said Marcia as she looked towards all the elders who were nodding in agreement.

"Of course!" said Afa, surprised again to be interrupted by Marcia. Yet she was also intrigued by Marcia's command of the room. She willed herself not to anticipate the "yes" or the "no" and instead she attempted to be completely present as Marcia spoke, trying to remain hopeful that there could still be a favorable outcome given her thorough presentation.

"Ms. Afa. We live on a small patch; however it is the biggest patch in the Amazon. Although so much of the South American rainforest has dried up, you can find some of the world's most astonishing natural wonders here in our patch. The Amazon River – the most voluminous river in the world – runs across its northern portion, originating in the Andes Mountains, the world's largest mountain range. This ecosystem was once home to more than 3 million species of animals and over 2,500 tree species. Much of the biodiversity of the rainforest is still intact in the patches. We, the Macushi, and our indigenous brothers and sisters in the other countries in South America, have always been here. We are the original inhabitants. We understand this forest better than any outsider, any scientist. We know its value, and we listen to the forest and water spirits to guide our decisions. We have withstood four hundred years of foreigners descending on Indigenous Peoples and their lands; the Spanish, Portuguese, English, and Dutch colonized us. We have a long history of outsiders making decisions about what is best for us. These foreigners who have oppressed us are what we call Kanaima, an evil spirit that takes the shape of an "outsider" who brings death and misfortune to us. Besides the Europeans, Africans were brought here as slaves, and more recently immigrants and their descendants have created a culture that is continuously imposed on us. But here we are, still standing, with our beliefs intact. Our rich culture and traditions are our own. Our beliefs about the spirits that govern our communities, and our land, are passed down from one generation to the next. You may call it "folklore" or "stories," but these beliefs and values are who we are. Now more than ever, we must fight to preserve our beliefs.

"Having said all this, we fear that the chemicals used in the rain making process will hurt our spirits, which means it will bring hurt, disease, and death to our community, as well as all the ecotourists that come to our land. The Macushi forest is a cultural forest. We are not the owners; we live within the forest." Marcia's eyes brightened and she raised her hands up to refer to the forest beyond the walls of the building. "We do not domesticate our forest; we become familiar with it. We are a part of the forest where spirits have lived before there was human life. Our failure to maintain relations with the spirits, the masters, and true owners of the forest may result in illness, death, or loss of resources. It is our job to take care of the forest, to mind it, to look after it – and we feel that the use of these chemicals would mean that we are neglecting our responsibilities to the forest and the spirits who live within it. We know that trace amounts of silver iodide are found in the water where the chemicals are shot out of the rockets to make rain. We cannot afford to hurt our water guardian spirits and anger the evil spirits which we call the oma. We must listen," Marcia concluded, with her hands over her heart.

Afa was silent, deferential. The community in the room likewise had their heads bowed, revering Marcia's words.

In the silent room, Afa walked over and took Marcia's hand. "I heard you. Thank you for giving me that perspective. I have one question." Marcia nodded. "The water guardian spirits you speak of – are they the Water Mamas?"

Marcia tipped her head up in surprise. "You know about our Water Mamas?"

"Vishnu told me a little bit about them after I met with you the first time."

"Yes, Vishnu respects our spirits, and he is helping us keep our culture alive for future generations."

"And I also want to keep your culture alive. I really do! The more I learn about it, the more I respect it. I don't believe that traces of this chemical will affect the composition of the water, I believe it won't affect the Water Mamas. It is such a small amount of silver iodide, it doesn't make a harmful impact on any ecosystem. It is being used already in many parts of the world!" said Afa emphatically.

"We have scientist friends in Australia and Canada who we have spoken to about the chemicals, who have informed our decision. In the Macushi language, Carib, the verbs "to know" and "to feel" are the same word. We feel and know that this rain program would not be good for our water," said Marcia.

"I hear that you have relationships with certain scientists who have told you the chemicals are harmful, but there are other scientists that you could speak to who would tell you something different. The scientists we work with..."

Marcia cut her off again. "The botanist, who is also our shaman, or what we call the Piaimen, has a relationship with the spirits and he can speak to them. The spirits have told him that the chemicals would hurt the Water Mamas," said Marcia with the absolute confidence of a true leader. Again, the other elders were all nodding their heads in agreement.

"While we appreciate all the effort that you and the United Nations have made to ensure our participation, we do not give our consent to the rain agreement," said Marcia. Her response was swift, cutting through all the platitudes and niceties, cutting Afa's agenda off at the knees.

Afa felt as if her stomach had dropped to her feet and all the blood drained from her face. Time seemed to slow down to the consistency of molasses, and she stammered to find a way forward. Deep down, Afa believed that the Macushi needed the program to survive, but she had not been successful in convincing them that the silver iodide would not affect the Water Mamas. She had to accept their final decision, knowing it was a decision that would affect the whole planet, a decision that she could not influence no matter how hard she tried. She had no choice. This was their right.

Marcia gestured with her arm for the elders to stand. "Ms. Afa, we see you as a friend of the Macushi. You are welcome on our compound whenever you would like to visit. We understand that you might be disappointed, we understand that we might be disappointing our indigenous brothers and sisters in other patches who have agreed to the program. But we must make the decision that protects us, as we see it."

"I understand. I really do." Afa sensed that pushing any further would be betraying them and she could not fight their decision anymore. It was over.

"We appreciate you, Afa," said Marcia.

"Thank you, I know. I will leave tonight to go back to Georgetown with Noel." She was staying at Noel's house before her flight back to New York, which was a relief. She didn't feel right staying with the Macushi after their decision. She knew she was welcome but also understood when her welcome was worn out. And it was time for her to leave. Afa thanked everyone, shook every hand, and Marcia gave her a hug. Walking back to the science station, a heaviness like lead overcame her and all she wanted to do was lie down. Back in her room, on her bed that smelled of mildew, surrounded by mosquito netting, her stomach clenched with dread at the thought of the conversation she would have to have with Kwame. The conversation would surely be heated because he perceived the situation only through the eyes of science and thought about the issue as an environmentalist. She wondered if he would scold them, saying that they didn't know what was best for them. But truly, Afa believed there was nothing else she could have done. The Macushi saw themselves as caretakers of the forest, and allowing the rain program threatened that stewardship.

Hours later, Afa rose sluggishly from her bed and walked to dinner. The weight in her stomach and chest hadn't dissipated despite her rest. She crossed the clearing in the multi-layered forest to the round structure where she could see all the other scientists, ecotourists and Macushi staff. She was waiting in line with her plate when her eyes met Vishnu's across the room. He walked towards her with his eyes firmly fixed on hers, an expression she couldn't read on his face.

"Good evening Ms. Afa. You ok? You don't look so good." He smiled kindly at her, and she offered him a weak smile in return. "Ah," he said. "What happened at the meeting? The Macushi said no to the program."

"They said no," said Afa, quietly.

"Damnit. I'm so sorry for you. Do you want to talk about it?" asked Vishnu.

"Yes. But, please, let's talk about it after dinner. I don't want to talk about it here, in front of the Macushi," said Afa.

"Yes, of course, after dinner," said Vishnu.

Afa and Vishnu stood in line for dinner together. The line had about 15 ecotourists waiting to be served ahead of them. She took a stab at small talk. "Do you know most of the trees on the compound by now?" asked Afa. It was an inane question, as they'd covered this topic, but it was also safe territory.

"Oh, there are so many trees. No, I have not learned them all, but I know a few of every type. The mangroves are my favorite. As for the plants, I shadow the botanist, and he teaches me."

"The dinner we are having tonight – pepperpot – seems like it is served very frequently here."

"Macushi would say it is the best dish in Guyana. It is made with a special sauce that comes from the cassava grown on this patch."

"I remember you talked about cassava in the car ride."

"Yes, you may know it as yucca root. It is interchangeable, but in Guyana we call it cassava."

"Oh, yes, I had chicha in Ecuador made from yucca root."

"Was it good?"

"It was…okay, I guess, it tasted fermented. They played a trick on me and told me it was made from the old women chewing the root and spitting it out." She smiled at the memory.

Vishnu grinned. "I have always wanted to try Ecuadorian chicha. Maybe one day I will get to the Ecuador patch."

"I hope you go. I think that the Ecuadorian Kichwa tribes, the Shuar, the Ashuar, I do think they will all sign the agreement. They have sent me correspondence that they are most likely going to agree, we just have to make it formal," said Afa hopefully.

"What happens if some tribes sign on and others don't, can you still run the program on the continent?"

"I don't know because I am not a scientist. My job is more diplomatic. I don't know the technical details. The scientists would be able to discuss that."

"Well, there are tons of scientists sleeping here every night. I've spent hours talking to them, you can talk to them too. Do you want to know some gossip?"

"Why not?" said Afa. Anything besides the meeting.

"There is one environmental science professor here that comes every year with his students, and he is in love with a young Macushi woman, Joanna. He wants her to invite him to come live with the Macushi people, but she doesn't have quite the same feelings for him. She likes him, but not as much as he likes her. He wants to marry her. So everyone kind of knows their story and everyone watches him pine after Joanna. It's a local drama."

Afa laughed and covered her mouth. "Really? Have I met this scientist?"

Vishnu grinned. "I think so."

"Oh, so everyone knows everyone's business here, then, huh? Do you know all the scientists?"

"Well, I am the only Guyanese scientist here and I am here whenever I am not in Jamaica so I kind of know everyone." Vishnu's eyes were bright and expressive as he spoke. He smiled slyly. "There is this one scientist, he is German. He is bald and lives in a very large body. He claims to have a woman in every port waiting for him as he travels around the world. He doesn't look like the type of guy who has women waiting for him. It's bizarre how confident he is. Some of these male scientists who spend all their time traveling alone are a bit delusional. I think the female scientists are more balanced, more levelheaded."

"Geez, I never realized the scientists spend so much time alone."

"Yes, they do. I am alone a lot, always moving between Jamaica, Canada and Guyana, so it can get lonely for me," said Vishnu. Afa was caught off guard by his admission and wondered if he was looking for a companion.

"I understand," said Afa.

"What about you? Do you ever get lonely traveling all over the world? Do you have a partner waiting for you in New York?"

"No, no one is waiting. I'm single," said Afa, the vulnerability of sharing this detail of her life creating a tiny flush in her cheeks. Vishnu did not respond with words. He just gave her an inscrutable smile.

After being served their dinner at the end of the line, Vishnu said "Let's sit with the Macushi." Afa looked around for Marcia and, spotting her at the end of the table, made eye contact and smiled. Marcia returned the gesture with a wide smile of her own. Afa took comfort in Marcia's warmth. Today's outcome had left her shook; keeping the line between business and personal was tricky, especially with how deeply connected Afa was beginning to feel to the outcome.

Over dinner, Afa and Vishnu traded stories with the Macushi staff of what it was like to live in the patch with ecotourists. Afa was happy that the Macushi were including her in the conversation, but she suspected it was because they all admired and felt close to Vishnu. After dinner, Vishnu motioned for her to follow him to a corner of the open-air room to a few chairs by the wood railing overlooking the clearing in the jungle.

"Hey, Afa, I'm curious if you know about the Sarayacu case? The tribe that fought the government of Ecuador and big oil in the international courts for illegally drilling on their land... and won the case? It was so many years ago in Ecuador. Did you ever hear of the lawyer for the tribe, Augustin Leon? He is an amazing man, and you should speak to him before he retires."

Afa nodded. "I stayed in the Sarayacu territory on the Pastaza patch they share with other tribes in Ecuador. Just before staying with them, I read about their case in the Inter-American Court. Why do you think I should speak to their lawyer?"

"Because he spent his whole life fighting for Indigenous Peoples to have consent for mega projects that took place on their land. I thought it might be interesting for you because he set up international case law and the precedent that made it possible for them to have consent, and for the UN to be required to ask for consent, according to international human rights standards. Talk to him. He's...well, he's a legend. He is respected by all the

Indigenous Peoples in the Amazon. I met him one time. He has great energy. And he would understand you and your work."

"Like Marcia? Great energy?"

"Yes, like Marcia. He is very wise, and he has seen and witnessed a lot."

"Ok, thanks for the suggestion. I will seek him out."

"You absolutely should. He would be interested in the position that you are in. It's because of his case that all the Indigenous Peoples in the Amazon know their right to consent."

Afa smiled, taking Vishnu's point. "I'll get his information from you before I leave. Thanks again, for the recommendation."

"Well, Afa, I am going to meet with some of the other biologists, so this is probably the last time I will see you before you go back to Georgetown. Let me give you my cell phone number so we can be in touch when you leave. I would like to keep talking to you. And if you agree, I would like to see you again."

Afa took out her phone. "Yes, I'd like that, too."

Vishnu grinned and reached for Afa's phone, tapping his phone number into her contacts. "I've just sent myself a text from your phone, so I have your number, too. Text me when you get to Noel's."

"I will," said Afa. They hugged and kissed on the cheek, lingering a little bit, then Afa broke away and waved to him as she walked away.

Afa was ready to leave the compound. She had come to terms with the fact that she didn't want anything that the Macushi didn't want, even though deep down she believed the scientists working with the UN when they said the chemicals were harmless. Like she had discussed with Vishnu, she wanted to honor and preserve the Macushi perspective, even though she did not agree with them. They had decided, and she had accepted their position. She felt at peace.

An hour later, after saying her goodbyes to Macushi staff and biologists that she had met, Afa was in her room packing her bags. Noel tapped on her door.

"Are you ready for a bumpy eight hours in the car?" asked Noel.

"Almost," Afa hoisted one of her bags onto the bed and zipped it closed. "Just need to say goodbye to the frogs."

Noel laughed. "Okay, I'll meet you at the van in ten minutes. It's just us this time – no supplies so we can spread out a bit."

They settled into the empty van and the driver drove them out of the clearing into the thick dark brush of the rainforest.

"I will really miss it here," said Afa.

"You will miss it even more when the patch gets smaller," said Noel.

"Marcia told you, then?" asked Afa.

"Yes, Marcia spoke to me before she met with you. I knew the Macushi would not consent to the rain program. I think it's a mistake. As a scientist myself, I think that the benefits outweigh the costs. Restoring rain to the whole planet is more important than hurting spirits in my book, but I also get that they really fear the chemicals. I know what mainstream and reputable scientists have said about the safety of cloud seeding. It's just some fringe scientists who believe that the chemicals pose any risk. But unfortunately, they have scared the Macushi," said Noel with a sigh and a shrug.

Afa shook her head. "I had really hoped they would agree to the program as well. But I respect their decision, and I understand their perspective. I don't know what this means for other patches, other indigenous communities in the Amazon. Now they won't be able to treat the whole Amazon ecosystem with one large artificial rain program," said Afa.

"It means that their patches are in more danger of dying. We must address the entire Amazon rainforest as one contiguous ecosystem. Plain and simple," said Noel.

"Yes, I agree," nodded Afa.

"I blame the so-called scientists that were talking to them, who had influence over them. Science is not the "neutral" and "objective" system that some people think it is. Science is just a system of organized knowledge, which always has personal cultural bias," said Noel.

"Yeah, the worst is when a scientist won't acknowledge other modalities for understanding the world that are not science, like the cultural or mythological perspective of the world." Afa was anticipating the reaction of her boss, Kwame.

"Ms. Afa, I have an example of where the line between science and culture is blurred. In the 1760s, the Swedish botanist Carl Linnaeus believed that botany was too sexual for women to study as he described the flowers as having sex, complete with sexual parts, and so on. Because of him, women were not allowed to study botany in many parts of Europe. He was inserting social norms of gender of the time he lived in into the science. It's not the fault of the Macushi that science is not absolute, is always evolving, and there is always an agenda. The Macushi were just talking to alarmist scientists who scared them. I am so sorry, Afa. This is not a good outcome."

"It's hard for me to accept that I didn't get through to them. Most of the world's respected scientists have informed the UN that the chemicals are used in such small quantities, they are harmless. But I couldn't get them to see that," Afa felt her stomach turn in on itself as her peaceful acceptance of the outcome of today's meeting slipped away with her fear that she hadn't done a good enough job. Afa and Noel drove out of the wet rainforest patch and entered the dry dusty savannahs of dried trees and stumps where decades before, humid jungles once stood tall. Afa sighed, then put her head back and slept for the majority of the van ride.

The yellow van finally pulled up to Noel's house in a gated community where no one ever sat in the guard house, according to Noel. As they opened the van doors, the deafening sound of the cicada insects filled Afa's ears. Noel's house was not big, and not small. It had two floors; all the interiors were made of a reddish color wood. There were wooden fans beating the air all throughout the house.

"Come in and get comfortable. Can I get you something to drink? Rum and coconut water? Water? Juice?" asked Noel. She flipped on some music on a tiny, turquoise radio that sat in her kitchen window and Prince was singing about partying like it was 1999. "Oooh," Noel said. "I love Prince." She rocked her hips side to side as she pulled some glasses from the cupboard and opened the tiny fridge. Her kitchen had green tiles on the floor, and there were bottles of rum on the wooden counter. The kitchen table chairs were sandy colored wicker, and looked as though they came from the patio of a Caribbean island.

"I'll have some juice," said Afa. Just then, she felt her phone buzz. She looked down and saw that she had a message from Kwame.

"How did the meeting go?" wrote Kwame in his text. No way was she explaining anything about the meetings with the Macushi over text.

Afa tapped out a quick response. "Let's schedule a one-on-one check-in when I get back to NYC. My flight is tomorrow morning."

Kwame responded with a simple, "Sure thing."

Noel brought over the juice and sat down next to Afa at the small table, fanning herself with a kitchen towel, still dancing with her shoulders, even as she sat.

"So I saw you were spending time with Vishnu. He's great, isn't he?"

"Yes, he's very knowledgeable," agreed Afa, worried that her attraction was obvious.

"His fiancée is supposed to come and stay with us at the science station for a week next month," said Noel.

Afa went pale as her stomach dropped and her face burned with heat. "His…fiancée?"

"Yes. Vishnu is engaged to a wealthy Indo-Guyanese woman from a very respected family in Georgetown. The wedding is in a few months. I'm going to the wedding."

Afa sat in disbelief. Vishnu was engaged? He was engaged and he had asked to kiss her!

"Vishnu is very good with the ladies. He is always flirting with someone at the compound, quite the ladies' man, so charming and women like him. It makes sense. I'm sure he is getting up to something with some of the other loose women at the station. He seems so passionate and sensual. But yes, he is engaged," said Noel. She swirled her towel around in the air like a lasso"...party like it's nineteen-ninety-nine!"

Afa felt so foolish for being affected by all his charm, his act, which is clearly what it was. But most of all, she just felt angry. She was red hot angry at Vishnu, and she was angry at herself for feeling special. He was engaged to be married, and he was getting into a hammock with her! And maybe there were other women too? She didn't know what to say, but she didn't want Noel to realize that she was just another woman that had fallen for Vishnu's swagger. Or that she was "loose," as Noel had described, and not professional. Afa stood and walked to the sink, dumping her juice and grabbing the rum that was sitting on the counter. She poured two fingers full and drank it, straight.

"Vishnu says his fiancée is a really skinny little thing. I haven't met her. Hopefully he can fatten her up! He likes girls that are what we Guyanese say, a 'thick slice.' Girls like you," said Noel, laughing and swatting at Afa's rear end with her kitchen towel. "Pour me some of that."

"How long have they been engaged?" asked Afa, her voice thick, pouring the rum into Noel's glass. She flushed to know she was his type, despite how disappointed she was. She swung as quickly to feeling angry at herself for caring.

"Oh, for about a year. He goes back and forth to Jamaica, so who knows – he could have another family there!" said Noel.

Afa was stunned, disappointed, embarrassed. He had even given her his phone number. He said he wanted to see her again. Why? What did he think was going to happen? What did she think was going to happen? She needed time to arrange her emotions, to gather her thoughts, so she changed the subject. She didn't want Noel to realize how affected she was by learning of Vishnu's engagement.

They moved from the kitchen to the couch, talking for about another hour, but Afa was distracted. Her mind kept pulling at the threads of every interaction she'd had with Vishnu, looking for signs that she'd misread things. When Afa yawned and said she wanted to get ready for bed, Noel showed her to the guest room, a small room at the top of the stairs that connected to an adjoining room by a little bathroom. The other room belonged to Noel's daughter who was staying at her dad's house, an Afro-Guyenese man who had become wealthy through real estate.

Afa wondered if he was paying Noel child support, but thought it was too personal to ask.

After laying in her bed with the ceiling fan beating for an hour, thinking about calling Vishnu, she turned on the light, and picked up her phone to text him.

"You are engaged? Why did you ask to kiss me?" wrote Afa, hoping her anger was obvious.

A few seconds later her phone buzzed.

"Because you are special," wrote Vishnu. This made Afa even angrier.

"How many special women are you kissing while you are engaged?" wrote Afa.

"Do you believe me when I say you are special?"

"What does it matter if I believe you or not?"

"I love your strength and your vision. We could work together in the jungle. Maybe we could be together, even. But in this life, I can't. My wedding date is set."

"Would you break the engagement for another woman?" wrote Afa. She could not believe she was writing such a thing, but she wanted to know.

"No. Our families have already blended financially. The marriage is a done deal," he replied. "But my wife and I have a special agreement. When I am working far from home, in the jungle or in Jaimaica, I can be with other gorgeous women like you. If she asks me about it, I tell her."

Afa stared at her phone. What sort of arrangement is this? What this the poly situation that Aliah talked about? While she was thinking, but her phone buzzed again.

"Maybe I can visit you in New York. We can still see each other...we can work together."

Afa couldn't believe what he was suggesting. Was he insane? Delusional? She shook her head and tapped out, "What would your future wife think of that?"

"She wouldn't mind. She can sleep with other men, too. But so far she hasn't."

Afa didn't even know whether she believed him about the arrangement, but this was way more than she could take in at the moment.

Then a thought occurred to her.

"Have you kissed other women since you have been engaged?" asked Afa

"Yes," he responded immediately.

Afa put her phone down and didn't hear it buzz again. Vishnu had stopped texting her as well. She laid in the bed feeling angry about Vishnu and nauseous at the thought of telling Kwame about the meeting when she got back to New York. Sleep finally came around dawn.

The next morning, after the few hours of sleep she'd managed, Afa graciously thanked Noel for having her as a guest. She called a taxi and amongst the roaring sound of the cicadas, dripping with sweat mixed with sun block, eyeliner and insect repellant, went to the airport for her flight back to New York City.

FIVE

Back in New York, Afa was working at home, a red brick brownstone in Harlem. The predominantly middle-class neighborhood had changed over the years as Colombia students and tech workers had taken up most of the housing. It was almost unrecognizable from what it had been even 10 years ago. The buildings still had their old worn-out facades, but now they were filled with coffee shops charging 20 dollars for a latte. There were bars brimming with young AI money. These days, the population is mostly white, but only a generation ago, it was mostly Black.

In 2015, a Whole Foods showed up on 125th Street, juxtaposed to the Spanish Harlem Caribbean restaurants that sold ham hocks to cook with collard greens, yams, oxtail, and goat meat. Certain types of businesses had become obsolete. Gone were the rickety outdoor stalls that sold bronze jewelry, straw hats, and colorful African clothing and mud cloth. Can't stop gentrification, Afa thought, questioning for the hundredth time her decision to move to Harlem. Perhaps she herself was a gentrifier.

Her apartment was warm in contrast to the chill of the outdoor air. The snow on the sidewalk was icy, as the powder snow had melted and refrozen at night. She was listening to her hip hop playlist, music of a completely different time, a different era. Mos Def, Common, Talib Kwali. There were Matisse art prints on the walls and the colors in the apartment were nude, camel and white. The furniture was modern. There were plants everywhere sitting in clay ceramic pots. Magnets on her refrigerator looked like shells and starfish. Comfortable in her own space, she felt she would have more power and confidence in talking to her boss about something difficult. It was time for their scheduled check-in where they would debrief informally, before she typed up her report for him to read. The phone finally rang, and she turned the music down, drew a deep, galvanizing breath, and answered.

"Hello, Kwame."

"Hello, Afa! Hopefully you have rested and recovered from your trip to the Amazon."

"Yes, I have. Thank you."

"Glad to hear it. I have another meeting shortly, so I hope you don't mind if we get right to business. How did the Guyana meeting go?"

"The Macushi don't want the program. They will not sign the agreement," Afa ripped the bandage, bracing for his reaction.

A pause. "What? Why?" He was incredulous. Clearly, he'd expected her to report that they'd signed.

"Because of the silver iodide."

"So – bring me up to speed. How did we get here – what reasons did they give exactly?"

"They were most concerned about it getting into the water. And they're going off reports from science they seem to trust."

"They didn't understand the program the way you explained it, so perhaps you didn't illustrate the facts, the benefits of the program to them – to the world! Do they know they are putting the planet at risk of not recovering our global hydrological cycle? It's not just about them, damn it! Did you explain it to them in a way so that they had all the information? Did you emphasize all the benefits of the program?"

In response to Kwame's heated words, Afa's back went up and she felt her own heat rising. The insinuation that she had not done her job well made her angry, displacing the self-doubt she had felt in earlier incantations of these conversations. He doesn't understand what my job entails, she thought. She knew he'd be disappointed but his strong reaction and immediate rush to blame her was insulting and she wouldn't stand for it.

"Of course I explained the benefits of the program fully and completely!" Afa tried to temper her response, but she felt defensive. She wanted to argue that she had done a good job, no longer fighting her own securities. Hell no. This time, she felt she had done all she could. She drew a deep breath and closed her eyes, gathering her thoughts.

"If you really explained the program, it should be a slam dunk! They would know the silver iodide is not harmful."

"But there are other considerations. They are worried about hurting their spirits that they believe live in the water."

"WHAT?! Did I just hear you right? Spirits told them not to sign the program?"

"No, spirits didn't tell them, but in their culture…"

"Do they know that the rivers will dry up, the forest will die and so all their spirits will die alongside them? If there are no indigenous people left to believe in spirits, then there are no more spirits." Kwame gave a rueful laugh. "Hearing this talk about spirits, I am left wondering if the Macushi understand how much they need the program. This is a huge mistake."

“They do understand, Kwame, they do understand all the benefits and disadvantages of the program. I don’t think you understand the importance of the water spirits to them.”

“Well, I don’t know if the effort you made was acceptable, Afa. You have not gotten consent yet. You will just have to go back until you get consent. We need Guyana to come to the realization that this silver thing has already been disproven. It’s an absolute imperative. We have to make sure that they see that the silver iodide does nothing, it’s an insignificant part of the process. The benefits completely outweigh the costs by a thousand percent.”

“I can’t go back! I mean, I won’t go back!” Afa was in disbelief that he would ask such a thing of her. “I did everything possible! And I did explain everything! I don’t believe in bullying them and cajoling them into consent. They made an informed decision!” She had lost any ability to temper her response to this request.

“What are their sources of information? You, or the spirits?” said Kwame in a way that seemed to demean all her efforts.

“You may not understand their cultural underpinnings that are not scientific, but nevertheless, that’s the source of their concern. Plus, they have fringe scientists telling them the chemicals are harmful – and those are the scientists that have got their ear,” said Afa.

“But that’s misinformation, Afa! It’s your job to educate them, to put out the correct information, to dispute the crackpot scientists out there pedaling these conspiracy theories about chemicals, rather than listen to the UN. I mean, who are they speaking to? Are these even real scientists?”

Afa realized that he thought her job was simple. Just say the facts and the correct science and get consent, without cultural considerations. The one thing she did agree with him about was the alarmist scientists talking to the Macushi.

"I will grant you one thing; I agree with you that they were persuaded by bad science."

"Well, if I can be frank with you, Afa, I am very disappointed. I am wondering if you were expert enough in the rain program to present it to the tribes. I can't help but wonder if I was the one doing the presentations, we might have a different outcome. They would have all the information, I would have shown them the benefits, I would have convinced them. I would have disproved their fears of the chemicals with facts. They are being too emotional, letting these stories about spirits inform them when rationality should have prevailed," said Kwame.

"I must respectfully disagree! You don't even accept that their culture plays an important role in who they are, and therefore, their decision. And please don't assume that just because you're a technical environmental scientist that you would have been able to convince them with more authority." Afa had completely lost her cool.

"This is ridiculous. They are just one culture. Do they know that their decision is affecting the whole world? The rainforests belong to everyone, not just Indigenous Peoples. Not just a few thousand people who live in the remaining patches. We all breathe the oxygen that comes from the rainforests, not just them. The entire global atmosphere benefits from rain in the Amazon. Were you able to tell them that? Do they understand the global ramifications? Without their consent, we cannot create rain for the entire Amazon. This is about the health of the planet, which we can't afford to endanger by listening to the perspectives and opinions of indigenous communities who are being informed by spirits. Therefore, I think the spirits lack the proper information. You have a job to do on behalf of the global community, of people whose lives would be better off if they had rain. You are here representing humanity, getting consent for something that would improve the life of everyone on earth, without fail."

"And I explained that this will affect global climate. Of course I did."

"And that didn't make an impression on them? There is a copious amount of information on how the Amazon wide rain program would mitigate global climate. I think you may need to push harder and let them know the global ramifications of their decision."

"With all due respect, Kwame, there is a fine line between me pressuring them to sign the agreement and asking for consent. Right now, I am trying to have them arrive at consent through the education I provide. They know they have the right to not sign the agreement, the right to decline the program. They know that they did not always have these rights. Now that it is a human right, they have been given the chance to deliberate on whether a program run by outsiders is good for their land. And after deliberation with the whole community, their answer is..."

"But that doesn't make any sense!" Kwame cut in, his anger and passion evident. "We are not like the other outsiders, we are not the miners, the loggers, the fossil fuel extractors. We are the UN! We are trying to do good for them, for their patch, for the rest of the Amazon. They are basically choosing suicide and taking the whole planet down with them!"

"We are not the only ones purporting to do good, to improve their lives. In my research, I read about a case in the 1970s, where a large baby formula company sent their employees out to the Brazilian Amazon, near the capital city of Manaus, dressed in nurse uniforms. The so-called "nurses" were there to tell the tribes about the benefits of formula, promising that the formula was better than breast milk, which is what they were all using to nurse at the time. These fake nurses told them they could buy the formula in the capital city on their trading routes, they told them their babies would grow big and strong. Eventually the baby formula employees were successful in convincing them, so the tribe began to buy the formula. However, to save money, they began to mix it with water, dilute it so heavily, and often with contaminated water. Soon, many of the babies developed diarrhea and died. After a few deaths, that tribe never accepted help from a foreign agency again. If they had been fooled by nurses, who else had an agenda to profit from them?"

"I understand that, Afa, but if we can show them that we are not bad nurses, that we will improve their lives… The UN wants to help them. And if they understand that it is for the good of the whole planet, they will sign the agreement. That is your job to make sure they understand that."

Afa could hear Kwame's passion, and she understood why he felt this way. She, too, wanted the program for all the tribes. But she also felt that Kwame was not hearing her. He kept putting her in the position of defending the way she did her work, and it was making her angry that he was trying to perforate her method, questioning her ability. She slowed herself down and tried again.

"Kwame. Pressuring them and forcing them into a decision isn't the way of the UN. We are not "helping" them if they don't want our help. We have to try to understand their reasoning for their position, even if we don't like it. We have to work with them. The Macushi understand the situation." Afa was no longer able to keep her own emotions out of her voice. "You're only seeing it from your perspective, Kwame, not theirs!" Afa felt stuck. He wasn't hearing her at all. He had no understanding of how culture, and fear of endangering the spirit world would prevent the Macushi from moving ahead with the program. His callous attitude towards the concerns of the tribes, the more forcefully he spoke to her, the more she empathized with the tribes.

By not even attempting to understand their perspective, she felt she herself was being misunderstood too. She wanted them to arrive at consent on their own and Kwame could not see that. He was not going to hear her. "They have their own point of view, and I did my best to understand it. I also did my best to make sure they understood me. That was my job, and I did it."

"Their point of view is misinformed! How many times do I have to tell you?"

"Kwame, we are going in circles. They made their choice! It is their right to decline the agreement!"

"What? Why are you defending a tribe that doesn't want the program? Why aren't you defending the global effort to stave back climate change, the climate crisis?" asked Kwame, his voice rising again.

"Do you even care about the Macushi? About the people that we are bringing the rain back for?" Afa surprised herself with how the volume of her voice had risen to meet Kwame's. It was unlike her to be confrontational, particularly with a superior, but she was getting sick of this circular argument.

"What kind of question is that? I do care about the Macushi! I want them to have their patch intact! But there are only a few thousand people left in the Amazon and I must think of everyone, including the farmer in drought ridden California, and Africa, and Asia! We are all connected ecologically, and you know that, Afa!"

"But you are so blinded by your science ideology! You can't see the human perspective!"

"You can call me ideological if I believe in my work, if I am my work. And I can question your dedication to your job, Afa, whether you believe in what you are doing. I am concerned that you don't care about bringing rain back to the global rainforests. With your failure to convince them of the benefits of the rain program, you have failed the global community," said Kwame, his voice strained and agitated.

Kwame's last statement stung Afa. But it also buoyed her, and she was not going to let him get away with it.

"How many real chances have Indigenous Peoples had to make decisions that affect them and their livelihood? They have been caretakers, good stewards of their ecosystems for centuries while powerful outsider countries have made decisions that have been bad for the earth and the rainforest? When have they ever had any say? Now they say something that goes against the global scientists, and they are the villain?"

"Villian or no villain, the stakes are too high for us to make a wrong decision right now. Rain doesn't fall naturally on most of planet earth. We schedule rain, for god's sake! They live in one of the last places where there is still natural rain and they have concerns that are unfounded."

Kwame paused, and there was silence on the line for a moment, each of them just breathing. She considered telling him if he really thought he could do better, he could go and try his hand at convincing the Macushi himself. Then he could come back with his tail between his legs when they denied him as well. But the last thing she would do was send a bully to their community.

She heard Kwame draw a deep breath. "Ok, Afa, we will circle back to Guyana later, as we have obviously reached an impasse. Let's just...let's put our focus on Brazil. We will put all our energy and resources there. There are two tribes in Brazil that you are scheduled to meet with, and perhaps they won't let the "spirits" get in the way."

Afa was relieved – and surprised – he had stopped. "Okay, great, Kwame. Thank you. I understand. I will focus on Brazil."

His words had pierced her confidence initially – the anticipation of his reaction had weighed on her so. But the conversation had served to make her stronger and surer of her abilities, perspectives and understanding. Afa's vantage point had changed. She knew that at the end of the day her responsibility was to the tribes, to advocate for them and honor the fact that they had a choice, to protect their free will. She was sure she had not failed in this regard, and that belief had given her courage. She'd held her own. Kwame could not sow seeds of doubt in her mind over how she had done her job, or how she would continue to conduct herself. She had anticipated that Kwame would not understand the position she was in, and she had readied herself.

But the weight of the world was on her shoulders – almost literally. This program was so vital to the health of the world, and it was up to her and her alone to convince all the Indigenous Peoples in the Amazon to sign on. And Kwame was right, in that way she had failed.

"I'll be sending you an email soon just to summarize our direction moving forward," said Kwame. "I have to go; I have another meeting."

A couple hours later, Afa received a follow up email from Kwame.

"Dear Afa, following our meeting, here are some action items:

1) Schedule a meeting with COICA to get a sense of where the Brazilian tribes are at RE rain program and plan your trip to Brazil.

2) Prepare to dispute the claims about silver iodide and present again the benefits of the rain program in Brazil.

Here at the Climate Modification department, we want the tribes to give their consent, to sign the agreement. However, we acknowledge that the UN guarantees Indigenous Peoples their human rights which includes consent of activities taking place on their land. I understand your job might be difficult in balancing these priorities, but I trust you will do your job well. I look forward to reading your report."

Pressure began to build behind Afa's eyes, and a heaviness settled in her chest. The job ahead of her seemed overwhelming, daunting. But she could not afford to become paralyzed. She took a deep breath, ran her fingers through her hair and the nape of her neck, held her face with her hands, and walked away from the computer.

***

Afa was waiting for Aliah in the blustery cold wind on the steps in front of the Met, the iconic museum they had chosen for today. The streets were lined with barely-there, gray-colored snow piled in the corners of the streets, leftover from a late storm. Winter would soon end in New York. She missed the humid warm weather of the Amazon.

Just then a cab door flung open, and Aliah walked out in an olive-green puffy jacket, her black curly hair spilling out of her hat. She beamed and waved at her friend.

"Afa! Americana Tropicana! I have not seen you in so long! You have to tell me everything! I want details!"

"Aliah, girl, bring it in! I am so happy to see you." The two women embraced and rocked back and forth, thrilled to see one another. "But first thing, your worst nightmare happened in Guyana," said Afa.

"What? What is my worst nightmare? Weird bugs? It was a bug, oh my god, wasn't it?" said Aliah, covering her eyes.

"A parasite laid eggs in my toe!" said Afa.

Aliah made the motion of opening her mouth and gagging, complete with a gagging sound. "That is absolutely so disgusting! How can you recover from that? Never! You can't! You will always know there were eggs in your toe. I'm going to throw up."

Afa laughed, pleased that her friend had reacted exactly as expected. "But my toe is fine now...after turning black!" She giggled at Aliah's disgusted expression. "Anyways. I could not wait to share that tidbit with you. I knew you'd love it."

"Yeah, thanks for starting there, Afa," Aliah said with disgust. "And let it be known I will nevvvver be joining you in the Amazon jungle." She drew out the syllables on the word never.

"Let's go in. It's freezing," said Afa, laughing and grabbing her friend by the arm. A few minutes later, they were walking into a large open room with skylights filled with marble statues. As they meandered around the statues, Aliah peppered Afa with questions.

"So, what happened?" asked Aliah. "Tell me everything and leave out anything to do with bugs and parasites, please."

"It's complicated. Macushi said no to the program. Although I don't agree with their decision, I understand their cultural and spiritual reasons for turning it down. They fear that the chemicals we use will hurt their spirits and hurt them. Specifically, they believe in Water Mamas, mermaid-like spirits that live in the river that will harm them if they don't manage the ecosystem the way that the spirits want. So as a result, they said no so they could preserve their Water Mama spirits and their culture."

"Hmm. That sounds like you're in a tough position. I want Indigenous Peoples to be able to preserve their culture too, and I understand they have ways of knowing and understanding their world that we may not fully comprehend." Aliah was an educator, so Afa had expected this from her. "But," she went on, "I also want more global rain. I think we all do. I do like the idea of 'Water Mamas.' A maternal spirit connected to the earth, like mother nature? Only it sounds like they give her more respect, no? I am curious, how do they know what the spirits want from them?" asked Aliah.

"Oh, there is a shaman who communicates with the spirits and tells the community what it is that the spirits want."

"Wow. Well, I can see how the credentials of a shaman would be hard to accept for a Westerner. And a scientist, no less"

"Yeah, I know. And I get that; I really see the struggle here. I also had a crazy vivid dream about a Water Mama that scared me a bit. Like, it was so real, Aliah. I don't want to do anything that would hurt or anger the Water Mamas, if they are in fact there in the jungle. I know I sound far out right now," said Afa, her face sheepish.

"Well, you said it, not me. These spirits are folktales and myths. Every culture has them. I do like that the spirit that has so much power is a woman. That's refreshing," said Afa with a slight uplift in her voice.

"But Aliah, maybe it's not a folktale or myth. That dream felt so real. Maybe the Water Mama was coming into my dream to warn me about the rain program."

"Um, I highly doubt it, girl. It was just a dream! I wouldn't attach so much meaning to it," said Aliah as she squinted at Afa as if to say, seriously, let's come back down to earth.

"Yeah. I could never tell my boss about my dream. He would think I'm crazy. He's razor focused on my job – on me doing my job well so that I get the right outcome according to him – which is of course, consent. He's focused on global ramifications, and I get it. But he put all the blame on me for them denying consent and questioned whether I gave an expert

presentation. He even questioned whether I cared about climate change." Afa rubbed at her hairline and frowned. "To be honest, I do feel some… I don't know. Did I do a good enough job? I worry that maybe I didn't explain the program well. Maybe not well enough or they would have consented. And he fans the flames, making me question myself. But I stood my ground. I defended myself. I did the best job I could within the parameters I had."
In Aliah's silence, Afa waffled. She shook her head and sighed. "Despite standing up to my boss, I still struggle with feeling like a professional failure because they did not give consent."

"Well, just to point out the obvious… you went to the jungle and you got a response that your boss didn't like, so you have to deal with that. Remember when I had issues with the principal of my school? There were so many decisions I made about how I run the classroom that he disagreed with. That's part of doing work that you believe in. You gotta deal with the fallout of taking a stand."

"Yes, I know. I anticipated his reaction, which at this point, I mean, he's pretty predictable."

"There could also be a culture clash, Afa. Living in New York where so many people are implants, everyone's perspective is culturally informed. My principal came from a more formal culture, and he didn't relate to the instinctive and spontaneous decisions I made to govern my classroom. I believe in responding to my students' needs, customizing the curriculum. My principal wanted me to follow the guidelines very strictly. Maybe it's like that at your job. Maybe your boss comes from a culture that is more hierarchical. Maybe he just expects you to fall in line."

"Well, I'm definitely not falling in line. Now that I've talked to the tribes, I just don't want to be another international agency that pushes them around. I have empathy for them, I don't want to push them so hard. I know he doesn't approve of how I do my job, but he can't fire me for not getting consent. I've been at the UN for years and I've been a part of some significant projects. So I know he can't get rid of me."

"Okay, that's good. You know your job is safe."

"Yeah, I do.

"Good."

"My boss has strong opinions about my work, even though he will never understand the indigenous perspective. He is a technician, he doesn't understand that their belief is also their culture, which is how I see things now," said Afa, surprising herself with the strength and resoluteness of her statement.

"Afa, to give another perspective, maybe consider that you might more empathy for the position that your boss is in. It's not that I don't see your perspective, but I have to push back on you a little bit here, my friend. Your boss is doing his job. And he is right. Getting consent does have global implications. It's not just about those indigenous people. Maybe he doesn't have to understand their culture. Maybe science is more important than culture. I know me saying this is kind of hard to hear. I mean, consider this, I love all the cultures living in New York City, but I would not want one specific culture's spirit governing the city," said Aliah.

Afa paused, a little thrown off by Aliah defending her boss. "I guess I see your point," she forced herself to say, pushing past her disappointment that Aliah hadn't just taken her side outright. And yet, she trusted Aliah to be levelheaded and honest. That's what she loved about her.

She sighed. "Okay. Yes. You're right. He is doing his job. But my job depends on earning the tribes' trust. If they don't, there is no way for me to do my job. And Kwame doesn't seem to have empathy for my job," said Afa. "Without learning about them and their culture, specifically the Water Mamas, it's way less likely that they will give consent." She paused for a minute and then continued to walk through a room of sculpted, muscular male statues carved out of white stone. She wanted the strength of the statues to move through her, as if she could absorb their strength.

"When I am talking to the members of the tribe, I feel like I'm back on my home island. They don't know if things will get worse or stay the same. The UN is promising that their lives will get better, but they don't know who to believe. They're the ones taking a risk. I feel like I'm my mom, begging my dad to see that sea level rise is happening."

"Look, I know you still doubt yourself. You always swing wildly between trusting your instincts and then questioning yourself. But girl, by now you know this story. Isn't that just from your father criticizing you so much as a child? You have shared with me how critical he was." Aliah clasped her hands, getting down to business, and Afa smirked at her friend. This was classic Aliah, putting on her teacher-who-took-developmental-psych-courses hat. "We have a lot of critical parents who constantly put their kids down, causing them to have less confidence. Parental criticism causes self-doubt. I have told you before. That's all this is. I think you have no actual reason be so critical of yourself and question your abilities so much, other than the seeds your father sewed in you. I know you are doing a great job, no matter the result."

Afa sighed. She had to admit some of what Aliah was saying rang true; it was just so hard to deprogram this stuff. "Thanks, girl, for going so deep with me on this."

"But this is so deep. All we do in life is make meaning out of events that have occurred, things people have said to us. We are meaning makers! We get programmed in our childhoods.

And when we become adults, we have to de-code, unravel the narratives that are holding us back. Like for instance, my narrative: I am not worthy unless I achieve success. I've been working on that for years. And you have your narrative, your underlying current of emotions that make you question yourself professionally, make you doubt yourself."

Afa nodded and gave Aliah's hand a squeeze of appreciation for her vote of confidence. She appreciated Aliah's breakdown of her persistent and nagging self-doubt. "Yeah, like you said, I am putting in all my energy. My motivation for working at the UN is helping restore the planet as climate change affects people's lives, just like it affected mine. I was affected by sea level rise; the indigenous people are affected by a lack of rain. The whole planet is affected by the lack of rain in the Amazon. We need to restore the global water cycle. I realize I won't get consent from the Macushi if I approach it just from the angle of science, so I am trying to find another way to achieve success with that tribe. I don't know how yet… I'm not prepared to strong arm them like so many colonizers, governments, and corporations have done before me. They have historically been oppressed, dispossessed, and this is their chance to exercise their own free will over matters that concern their land," said Afa.

"So…the question is, should a small group people, albeit a historically oppressed people, who have been beaten down for centuries, have the power to decide what is best for the entire planet?" asked Aliah, making it all so simple.

"Well, yes. That is the question, and it's not so simple." Afa inhaled and sighed deeply.

"You will do the right thing Afa, you will figure it out," assured Aliah. "You wanna hit the cafe now? I'm hungry." And just like that, she'd pulled Afa out of the jungle and back into the city. But only momentarily.

Once they were seated, and Afa had her tarte and Aliah her quiche du jour, Afa dropped the more tantalizing news of the day.

"So, also, when I was in Guyana I met someone. A colleague, and someone I really respect. But… I almost kissed him in the jungle."

"What? You kissed someone? Who? When? Tell me more!"

"I said we almost kissed, and you act like I haven't kissed anyone in years." Afa noticed Aliah's "no lies detected" expression, and she rolled her eyes at her friend. "Alright, fine. But he asked if he could kiss me, I said no."

"What? Why…" Aliah started, but Afa held her hand up.

"...because of our working relationship. I thought it would make things awkward, and I didn't want the scientists and others at the station judging me."

Aliah nodded, saying, "Okay. Fair. I get that. Continue."

"But then…" Afa dragged this part out for dramatic effect. "Then, I found out later that he's engaged to be married. He didn't tell me himself. He kept it from me." She waited for Aliah's reaction, and her friend did not disappoint her. Aliah's eyes grew wide, and her jaw dropped open. She leaned her head forward and grabbed Afa's arm.

"What. The actual. F. He's engaged and didn't tell you? And he asked to kiss you?"

"Yep. So, I'm glad we didn't kiss, you know?" said Afa in a confessional and quiet tone. "But I have so much respect for him, even though he is also a ladies' man, and-"

"Hold on? Respect? Like, on a professional level?"

"Well, yes, absolutely. I felt that he really understood my work and what I was doing. And he was someone I could lean on for support while I was there. Plus, his work was so interesting. It was like he understood me, you know? And he was so charming and such a good dancer." Afa, drifted off, thinking how wonderful that night was when they danced so close.

"Ummm, Girl. Wait just a minute. Good dancer?" asked Aliah, her eyes sparkling, a grin tugging at the corners of her mouth.

"No, well, I mean, that isn't important." Afa waved her hand to dismiss the detail. "He was so knowledgeable about the Macushi people, their culture, he cared about them so much. He was also very comfortable in the forest. Like, he picked up snakes when we were walking. He was just a man of the earth! Working with the Indigenous Peoples, trying to understand how they manage the forest. It was a mental connection we had, and I just liked his approach."

"Mmmhmm." Aliah was still grinning. "What's his name?"

"Vishnu."

"Vishnu? Is that an Indigenous name? I mean it's not like I know all the indigenous names but..."

"Yeah, his family is Indo Guyanese. There are a lot of Southeast Asians that have settled in Guyana. Anyways, he is a biologist who works with Indigenous Peoples. And he understood the situation I was in at the time," said Afa.

Aliah leaned back in her seat and took a long drink from her glass. She studied her friend, then licked her lips. "Wow...so what now?"

"What? No. Nothing! Nothing even happened..." Afa started. "Well, we laid in a hammock..."

"Together." Aliah said it as a statement.

"Well, yes. That's where he asked to kiss me. We were close." She looked at Aliah, who was smirking. "Look, I'm not denying that there was attraction. There was. I hoped maybe something would come of it but…"

"Afa. Just listening to you go on and on about this Vishnu, it's clear you really liked him."

Afa sighed. Aliah was right. But so what?

"So, real talk. What now? Where'd you leave it?" Aliah asked.

"Well. Nowhere. Nothing romantic. He's engaged. But I am open to working with him in the future."

"I dunno Afa, although I was dramatic about him being engaged and not telling you a minute ago, I've had a change of heart. I do think you could have kissed him. I mean, being engaged isn't the same as being married."

"Isn't it though? And anyway, he's locked into it. We discussed it. He's got some financial deal with her family or something, I don't know all the details, but it sounds like it's for real." For some reason she couldn't quite identify, Afa held back from sharing with Aliah the fact that Vishnu had hinted that there was a possibility for more between them despite his impending marriage. She wanted to keep that a little close to the chest for now. Aliah was saying, "I mean, it's been a while for you. Your hymen might have grown back over! You haven't been with a man for – what? How long has it been since you were seeing Cyrus?"

Cyrus was a man that Afa had met on the subway when her bag of groceries broke and the contents spilled all over the subway car. He had quickly stood and helped her chase her lettuce and oranges down the sticky subway car floor as they rolled under people's feet. He had walked her home holding her groceries in his bare hands and then he asked her out on a date. Afa liked Cyrus and had seen potential for a future with him; he was kind and thoughtful, always treated her with respect. He made a good living, was close to his family, and had a romantic streak. Like the time he whisked her off to Cape Cod for a weekend. But a few months into dating him, when she thought they'd take things in a more serious direction, Cyrus revealed that he wanted to marry an Iranian Jew like himself. He had told her he was open to dating her, but he couldn't end up with her. At the time, Afa was confused – should she continue to date him to pass the time, or set herself free to find someone who she could have a future with? In the end, the gnawing feeling that he would never choose her, no matter what she did, made her feel rejected. She'd ended things. And she hadn't dated anyone since.

"It's been a year since Cyrus," said Afa quietly.

"Yeah, and before that was Ivan, who told you he was in love with you and then disappeared, no calls, no texts ever again. Remember you grief-ate a whole cheesecake you were so upset? Oh, you were so upset about Ivan! It took you months to recover! I really had to help you after him," said Aliah.

"Yeah, I remember," admitted Afa begrudgingly. "And what exactly is the point of torturing me with this recap of my so-far-so-unsuccessful love life?"

Aliah winced. "I'm sorry." She picked up the check, shaking her head at Afa when she took out her wallet. "Let me get this. You got lunch last time. Anyways, I just think you should be open minded around the men you date. Think outside the box. You never know when the right one for you will appear. Maybe Vishnu and his fiancée are poly. But engagement aside, he sounds like he ticks a lot of boxes."

"Well, I will try to be more open minded," said Afa with a wink. "I know you are more open than me."

"It's not that I am more open minded, it's just that I'm out there. I'm dating more often than you are. I mean, yeah, I guess because I am polyamorous and don't see anything wrong with dating someone who has other partners, that makes me more open minded. Girl, it's so common. Almost everyone on the dating apps is poly. Welcome to the future."

"Yeah. I hear you. I just don't know if I am poly."

"Well, I think you should try it out. Monogamy is dead. It used to work when we lived to 40. But now we live till 100, and it doesn't make sense to have the same partner for all that time."

"Let me think about it. I promise I'll consider it."

"Okay girl. Consider it. Liberate yourself!"

Afa nodded vigorously and put her hand on Aliah's hand as if to say thank you.

***

Afa was sitting at her desk at the United Nations, typing feverishly on her keyboard, attacking keys with the plump part of her fingers. She was writing the final report for the rain program in Guyana, based on her conversations and her analysis. Because she had not received consent from the tribes on the Guyanese patch, she was recommending that the rain program would have to avoid that geographic region, which meant that the South American rainforest could not be treated like one contiguous ecosystem by the rain rockets. She didn't know how the planners and scientists would avoid one tribe's land, or if it was even possible, but her heart hurt thinking that their patch would not receive rain when all the neighboring patches would. As she was finishing her recommendation, her colleague Nick walked in. Nick was a lawyer who specialized in sea level rise representing the Association of Small Island States. He lobbied the general assembly to take more protective measures to protect climate refugees who were leaving islands for other countries. Afa had bonded with Nick because of his work with the islands, and now they were both working in the rain programs.

"Hey Afa, I just wanted to say hi. I haven't seen you for a while," said Nick with a warm smile.

"Nick! It's nice to see you. I miss seeing each other all the time."

"I just heard that you're going to Brazil to talk to the communities there about the program."

Afa nodded "I am. Headed for the big patch."

"Yeah, yeah. That's a big deal. Biggest patch in Amazon country. I've always wanted to go to Brazil, but the crime in the cities is terrible. You be careful. When are you going?" he asked.

"I'm not sure exactly when I'm going, but soon." Afa had to work up the courage to visit other tribes that perhaps would not give consent to the program. She still felt emotionally exhausted from her last trip to Guyana.

"Okay let's touch base again when you know the dates. I want to hear your plans for visiting the communities. Maybe I can help. I have some capacity right now."

"Thanks. I don't think I need help so far, but I will let you know if I need your input. I might need your help going over something in my report.

"I'm happy to help you. Maybe we can go to coffee some time? I'd love to hang out with you outside of work. I'm impressed with the work you do. It takes a strong person to do what you're doing and keep at it the way you are."

"Aw, thanks Nick," cooed Afa. They had always been friendly colleagues, and she considered him a friend, but in this moment, Afa wondered if he was flirting with her. Or maybe he just sharing his observation. She wasn't sure. "What are you working on right now?"

"The Africa rain program, which is going really well. The rockets have been working on a schedule and we are already seeing some vegetation coming back – primary succession of the tropical rainforests. I don't know how long it will take for us to restore the ecosystem to the baseline year, but ecosystem succession has begun. Anyway, I can always hop on your project with the free time I have right now."

"I'm glad to hear Africa is going well! That's wonderful! I'll let you know about Brazil, for sure. And hey, thanks for stopping by. Sorry I can't talk more right now, I'm just in the middle of this report. But I'll touch base with you when I get to Brazil to let you know how it's going, maybe we can go for coffee after I get back," said Afa, as she turned back to her computer. There was a sense of responsibility she felt to all her colleagues and coworkers who were engineering the international agreements and programs, and she couldn't deny it: she wanted success like Nick. She wanted to feel accomplished when she talked to her colleagues, strong in the belief in herself, and proud of the job she had done.

SIX

"Welcome to Brazil! I am Paulo, like the Brazilian author, Paulo Coelho. Please, let me get that."

The middle-aged man with a disheveled appearance reached for her bag. "Is this all you have? Just one bag?" Paulo looked at her with a confused expression, his thick eyebrows furrowed, his caramel colored and calloused hands reaching for her small duffel bag to carry it to the car.

"Yes, I am here just for a few days...I didn't need that much."

The old man gave a curt nod. "Here in Sao Paulo, we have lots of crime unfortunately. You cannot put bags on the seat next to you or below your feet in the car. Men on motorcycles with guns will drive up to the car at the stoplight and demand everything that you have. We must put all bags in the trunk, your purse included."

"Okay, no problem," said Afa with a nod.

Paulo continued, "Good thing you have me as your driver because I can keep you safe. Do not use your cell phone in the street as people snatch them out of your hand. If you must use your phone, go into a café, restaurant or store, and even then, you have to be careful."

"Okay, I will be very careful," said Afa in her most dutiful voice. "Thank you, Paulo. The Alchemist was one of my favorite books, by the way."

Paulo smiled at her, the first time he had smiled since greeting her.

During the car ride, Afa was reviewing what she had read in her dossier to prepare for this trip. She was meeting with the Yanomami near the Venezuelan border, and the Kayapo located right in the middle of the country. The satellite images were difficult to read. Both patches were big but had major roads running through them which created erosion along the edges of the patch where the roads met the land. The first clearings that appear in the forest along the roads were always in a fishbone pattern referred to as "fishbone deforestation" and over time the fishbones collapsed. The roads brought gold prospectors in the 1980s, and miners made up five times the indigenous population that lived on the territory. The miners found gold, which led to more deforestation and destruction of indigenous ancestral land. The Brazilian government in 1992 began to recognize the territorial, cultural and civil-political rights of the Yanomami and Kayapo. However, the fires and the deforestation sped up in the 2000s until dieback began, when the scaled back forest began to self-destruct as a result of not being able to maintain the water cycle.

The Brazilian patches had the most contact with major cities in Brazil as Indigenous Peoples went back and forth between the city and the patches, making them most connected to the outside world. The Kayapo and the Yanomami had achieved self-autonomy, or what the UN calls "self-determination" by participating in the global economy on their own terms. The tribes had come a long way since the "slave-hunting" expeditions that started in 1654. There were many other tribes at one point in the forest, but as the forest began to disappear, the majority of tribes were displaced.

"Here we are." Paulo used his key to open the trunk of a rather rusty looking car. He placed her bag inside, then reached for her purse, looking at her for consent before taking it. She handed it to him with a nod and a smile. "You work at the United Nations headquarters in New York? That building was designed by a Brazilian architect, did you know that? Brazilians have done many great things all over the world!" Paulo beamed with pride.

"No, I didn't know that. How interesting!"

"Yes! Brazil has produced many great things, but there is a lot of poverty in Brazil too. Here in Sao Paulo, you can tell a person's wealth by their feet. Poor people have dirty toes sticking out of their sandals, and the rich all have clean, nice-looking toes."

"I see." Afa did not know what to make of his comment about the feet. She was aware of the gap between the haves and have nots in Brazil. It was one of the largest gaps in the whole world.

What Afa did notice looking out the window, and not in her prepared dossier, was a city of skyscrapers in contrast to nearby hillsides covered with slums of small makeshift houses. She had read that the favelas were neighborhoods of informal homes built against the steepest undesirable hillsides with makeshift plumbing, lighting, and waste removal. These neighborhoods were run by informal governments of gangs, or drug dealers, where police wouldn't enter. Favelas were occupied mostly by Afro-Brazilians, a reminder of slavery in Brazil.

As they were driving through the city of Sao Paulo to the hotel in Jardins, Afa checked her phone. Nick, her colleague at the UN, had texted: "Call when you can, we should talk."

This was a bit odd: Nick had offered to help her organize details, and they did share a common superior in the office, but he rarely reached out to her for a conversation when she was on a trip. She decided she would call him later after she arrived at the hotel.

A second text was from Vishnu, who said, "Are you in Brazil yet? I can come see you if you want. I'm thinking about you and miss you. Call me."

Her stomach did a tiny flip to know that he was thinking about her. The corners of her mouth lifted, but just as quickly, she frowned, not even wanting to admit to herself, really, that she'd been thinking of him too. She squeezed her eyes shut. She really didn't know what to make of Vishnu and his arrangement with his fiancée. How was he able to get away from her to come to Brazil? Would he tell his fiancée that he was traveling for work? Was this part of the "arrangement" he had mentioned? Was it an open relationship? How honest was he with his fiancée? Afa shook her head again. Never mind. It didn't matter anyway, because she was not going to ask him to come to Brazil, despite the temptation.

A third text was from Kwame who simply said, "I know you can do it. Good luck." Afa closed the text app, tapping her screen hard and pressing the side button to darken her screen. She resented the way Kwame managed her, and almost anything he said annoyed her.

It wasn't as if Afa had begun her work with the tribes with the attitude that they would have reasons to say no to the program. She, like Kwame, thought they would all say yes, of course. But after working with the Macushi in Guyana, she had arrived at a new place, one that included space for the feelings and beliefs and attitudes of the people she was working to get to know. She knew they may have their own reasons for the decisions they made, and she wanted them to know she understood – or at least respected – their reasons. Walking into the meetings in Brazil, she felt that she was working for the Indigenous Peoples, not the UN.

"This is the neighborhood of Jardins. One of the nicest neighborhoods in Sao Paulo," said Paulo, interrupting Afa's thoughts. She wondered what he was being paid as her driver. Then she wondered about his toes. The UN was paying for the driver and the hotel that the Indigenous Peoples were staying in, where they would meet in a few hours. As she looked out the window at the tree lined neighborhood of Jardins, she started to see armed guards with machine guns standing outside the entrances of buildings, hotels, and residences. And on the street, she could easily identify the service workers and domestic workers who were employed in the upscale apartments – their weathered black or brown skin and faded clothing setting them apart from the wealthy Brazilians being let into the buildings. Afa found the inequity in Sao Paulo to be very extreme on a visual level alone. She wondered how the Kayapo and Yanomami tribes felt about being in Sao Paulo.

When they finally arrived at the hotel, Paulo got out to get her bags from the trunk, the deep creases in his sunbaked skin deepening as he smiled at her, gesturing to the front door. She pulled out some money to tip him and he graciously held out both hands to accept the money. Armed guards stood on either side of the hotel doors, and Afa found this to be very jarring. She walked past the guards without making eye contact with them, and headed for the front desk to check in. The hotel's sterile interior, with its industrial soap and new carpet smell, stood in stark contrast to the city they had just traveled through. Afa would need the hour she had to shower and change her clothes before she met the Indigenous Peoples in the hotel's small conference room. She didn't know if an hour would be enough to put the pieces of this day in their place, but it was all she had.

Afa looked cheerier than she felt in her pink cotton skirt and white shirt, with no jewelry and her hair up in a messy bun. Her blue eyeliner would stay in place until she started sweating. She had steamed out the wrinkles by hanging her clothes in the humid bathroom while she showered. The weather in Sao Paulo was not too hot or too cold, but the hotel felt cool because of strong air conditioning. Afa picked up her laptop, and all the files and notes she needed to present to the tribes. The heavy metal doors of the elevator opened and stepping inside, Afa pressed a hand to her belly, taking a deep breath. She was incredibly nervous. How comfortable would the tribes be meeting her in a hotel conference room, she wondered? She had met the Ecuador tribes and the Guyana tribes in their own territories and villages, but meeting here, in this corporate and sterile environment, she was concerned was a bad plan. Then again, the Brazilian tribes were most integrated with the major cities, so perhaps they were comfortable in Sao Paulo; she shouldn't make assumptions of what their comfort level was, what was too "posh" or refined. Her nerves were really having their way with her.

Bracing herself, and drawing a deep breath, she stepped inside the conference room, and everyone in the room stood up. She smiled broadly and nodded at a few of the members of the group who bowed at her in return. From the photos in the dossier, she recognized the two Yanomami leaders by the red, black and white markings on their faces and the blue feathers in their hair. Across the room were the two Kayapo leaders wearing plain western clothes. Stepping to the lectern at the front of the room, she addressed everyone at once.

"Hello, everyone. Thank you so much for coming to meet with me, and thank you for coming to Sao Paulo. My name is Afan Sione, and I grew up on a South Pacific Island that went under the sea. Now, I am working on rain agreements at the United Nations. Shall we begin with introductions? Please, take your seats."

"Ms. Afa, we do not need to sit down. My name is Oscar of the Yanomami. This is another elder from our village named Shori." Oscar spoke for both men and looked very stern, his mouth a straight line.

This was a bad sign. Immediately Afa started to sweat through her shirt.

"Hello. My name is Joao, and I am Kayapo," said a man wearing a t-shirt, khaki pants and sandals. His face looked more relaxed, but focused.

"I am Adriana, and I am also Kayapo." Adriana was a young woman, younger than the men in the room. She was wearing a dress with sneakers. She was very pretty, with straight long black hair and delicate hands as she pulled her hair behind her ear.

The Kayapo leaders took their seats while the Yanomami stood straight up, erect, not flinching. The fact that they were still standing made Afa feel very tense, but they themselves did not seem tense – just serious. All the leaders spoke excellent English, consistent with what she had read, that the Brazilian tribes were the most globalized.

"Ok, well let's get started, shall we? I wanted to start by summarizing the rain program and its benefits, and then I thought I could discuss the agreement, what it would entail, and your role," said Afa as she shifted her eye contact between the Kayapo on one side of the room and the Yanomami on the other. Oscar, the Yanomami leader who had introduced himself and the elder, took a few steps forward.

"Ms. Afa, may we have a chance to speak now? Before you speak?" asked Oscar.

"Uh, yes, of course," said Afa, feeling startled, her stomach taking flight.

"We will not be poisoned!" said Oscar, raising his voice to fill the room.

"Umm... I understand your concern...but...Sir, I have not presented on the rain-" Afa stammered.

"No." Oscar cut Afa off as she was speaking. "We have heard of your program. The United Nations is not acting in our best interest!" Oscar was very angry, and his voice was forceful, as if he was pushing Afa into the corner of the room with his large presence. The other Indigenous Peoples in the room were looking at him as he spoke. The other Yanomami elder was nodding, his arms folded against his chest.

"I assure you that the United Nations takes the human rights of Indigenous Peoples very seriously and I'm here to-"

"Where was the United Nations when our forests were burning, being cleared, drilled for oil, and cut down for wood, when our animals were dying?" Oscar kept the same energy, filling the room with the sound of his voice.

"Yes, I understand your frustration, I really do...I...um..." Afa was stammering. She was not prepared for this level of anger from the Brazilian indigenous leaders. She was prepared for dissidence, but not like this.

"Where was the United Nations when our babies were dying of diarrhea? From contaminated water from chemicals used by the gold miners on our land. Our land is already contaminated, and you want to pour down more chemicals from the sky so that we have rain? What good is the rain if we are poisoned? Who is the rain for?" Oscar was speaking faster and was piercing the air with his words. He was breathing faster, and his cheeks had turned red.

"But the silver iodide has been shown to have very little effect on human health by qualified expert scientists, and I would love to present to you the information that-" Afa was speaking quickly to fit in as many words as she could before he cut her off again.

"No! The Yanomami say no to the rain program! We have come to say that we do not give our consent to the agreement! We don't trust you! What does the United Nations want with our land? We are angry! Why have you brought us here? We don't trust your intentions! What else do you want from us? The rest of the world has taken everything from us and has given us nothing in return. We say no!" said Oscar.

"Ok I understand, Oscar, I really do. I hear that you don't trust the program, but...could I please be allowed to present the science, and then continue the conversation?"

"No! Do you not respect my words that I have come here to say?"

"I do respect you! Yes, I hear your words. You don't want the program. Okay. What about the Kayapo? What do you say?" said Afa turning hopefully, tilting her head to the side when she addressed the Kayapo leaders sitting on the other side of the room.

"We agree with the Yanomami. The Kayapo are not interested in a rain program that would introduce more foreign chemicals to our land." The young Kayapo woman seemed much calmer than the Yanomami, and she was less emotional when she spoke. Her hands were folded neatly in her lap.

"I have heard both of you. I respect your words. But if I can have a minute to present to you the information from the UN on the silver iodide..." pleaded Afa, suspecting she was running out of chances to refute the fears of the chemical.

"We have heard enough! We have come here to say no. Tell the United Nations that we do not want to be a part of your rain program, nor any other type of program on our land!" said Oscar, his voice becoming harder with every sentence.

Grief began to rise in Afa's chest over not being able to even begin to present her case, and that they could not hear what she had to say, but she felt there was no point in pressing on. It was time to give up, to respect the wishes of the two tribes, as she had done with the Macushi. She did not have the space to explain any further. She had to put the responsibilities of her job aside and lead with her heart. Even if they did not understand the science behind the silver iodide, they were scared, and she could empathize with that. She remembered the fear that they felt on the island when the land started sinking into the ocean. It was not for her to argue or fight for the rain program if just the thought of it brought them so much pain.

"Alright. I understand that you have come to your decision. Thank you for coming to meet me. I will take what you said back to the United Nations. I am sorry that I may have caused you any discomfort." Afa's voice had become quiet. The weight of their decision, particularly without the benefit of her presentation, was crushing. In the deepest part of her heart, she did not want this outcome. Sweat beaded across her forehead as she started to pack up her things.

"Thank you, Ms. Afa. I hope you have a good flight back to New York." The young woman, Adriana, offered her a small smile before turning to go. Afa smiled weakly back at her.

With her computer and files in her hand, Afa left the room, crossed the cold, sterile lobby, and pushed the button for the elevator. To be on the receiving end of such angry, vehement comments had gutted her, and she began to cry very quietly. Silent tears rolled down her cheeks, and she drew a deep, shuddering breath

The last thing Afa wanted right now was to talk to anyone, but realizing she had told Nick she'd call him after her meeting, she resigned herself to the task as she stepped into her room. A friendly voice might not be the worst thing in the world. After all, they had been colleagues for 10 years, and he might offer a word of comfort over her defeat. He was always so interested in her and her work. She pushed send on the call and put him on speaker as she popped open a bottle of wine from her room's mini fridge. The UN can afford it, thought Afa to herself as she dialed Nick's number.

"Hello, Afa, thanks for calling." His tone was so serious. Afa thought, Oh, shit. What now?

"Hey, Nick. I just met with the Brazilians, and they don't want the rain program. They were adamantly opposed, and I am making the UN pay for expensive mini fridge wine. How are you," said Afa, her voice purposefully lifting and betraying her deflated mood. She threw herself into a chair and propped her feet up on the windowsill.

"Afa, that makes this harder, what I have to tell you."

"Why, what? What do you have to tell me?"

"I have been a part of some conversations here at the office. I'm so sorry, Afa, but I think they are going to go through with the program even if they don't have the consent of the Guyanese and Brazilian tribes. I know how hard you've been working on the ground..."

"WHAT?" Afa cut in, sitting straight up. "No, no, they cannot do that!" Afa felt that the air had been sucked out of her chest.

"I know...I know, it's not what we thought would happen when working on a UN agreement."

"But what about the Declaration of Human Rights of Indigenous Peoples? Are we just going to ignore the fact that the UN has granted them rights?"

"There is big money behind the decision to move forward. Of the countries who have a permanent seat on the UN Security Council – China, France, Russia – they have already invested so much money into rain technology... and they've been pushing. The Brazilian government found a loophole. The Indigenous Peoples have rights to the forest, but they don't have rights to the subsoil; the state government does, and they are claiming that the rain affects the subsoil. Brazil has decided to move forward, trumping the rights of the Indigenous Peoples. And Guyana is trumping the rights of the Guyanese, citing that the lack of rain is a national security issue."

"So the UN isn't going to listen to the Indigenous Peoples now. After all we have done." Again, tears welled in Afa's eyes, a lump in her throat. "I can't believe this. This can't be happening. All the work we have done so far." Afa began to sob, shaking her head in disbelief, a deep pain hardening in her chest. "This can't be happening, Nick. Are you sure they are going to move forward?"

"Yes, it's what everyone in the department is talking about. They're going to notify you when you get back tomorrow. But I wanted to catch you earlier and be the one to tell you. I knew this would be hard for you. I am so sorry, Afa. I know how seriously you take the voice of the community. I really respect you, and I couldn't let you find out that way."

Afa hung her head. "So, who is going to tell them that the program is going forward: me? Is that what they expect me to do now? Be the traitor?" said Afa, sobbing, unable to control her emotion.

"I don't know who is going to tell them, but I assume they're going to tell you that also when you get back."

Afa gulped in a breath of air, heartbroken over the relationships she had forged within those communities that surely would not survive this betrayal. Noel. Marcia. Her heart ached.

"Thanks so much for telling me, Nick. I have to get off the phone and get myself together. I appreciate you reaching out to me so much. I'll see you in the office."

"I'm sorry, Afa. I know how much you care." Nick paused, and Afa could sense his empathy through the phone line. "I'll see you soon."

A barrage of emotion spilled over inside Afa; she was furious, devastated, betrayed, bereft, confused. It felt like all her work had been for nothing. She was grateful that Nick told her; he was always so caring towards her. Her throat constricted, tight with despair, and her body ached with grief. When she thought about the Yanomami and Kayapo, recalling that she had left them with a promise that she would respect their decision, she cringed. She had stood in front of them and said she would honor their perspective.

There was only one person that she wanted to speak to, one person she knew would understand her. At that moment, Afa knew what she wanted. Her desire to be understood – for the Indigenous Peoples to be understood – was stronger than any other force. Who else valued their desires, the importance of their indigenous perspectives? It felt like no one understood better, cared more than him.

She picked up her phone and dialed him.

"Hello, Afa."

"Vishnu. I have very bad news."

"I can hear it in your voice. What's wrong?"

"The UN is going to go forward with the program without the consent of the Guyenese and Brazilian tribes! And like the Guyanese, the Brazilian tribes just said no. But instead of talking about how chemicals would hurt their spirits, the Brazillians have a general distrust of the UN, who, of course, just proved them right!" Afa was crying while she spoke, taking deep breaths in between sobs. "I can't believe this is happening! I just can't believe it!" She threw herself onto the bed.

"No! Afa! I can't believe it either. From everything you told me, I thought the UN was on the side of the Indigenous Peoples and protecting them."

"I know, I know." Afa had to pause and catch her breath. "I just spoke to my colleague who says there is big money pushing for it. Different countries have funded it and are pushing for it, and the governments of Brazil and Guyana said yes in spite of what the communities might wish."

"But what about the rights of the Indigenous Peoples within Brazil? Brazil has the strongest protections. Surely, they are protected by the state?"

"The Brazilian government has said they can make decisions regarding the subsoil land, despite the demarcation of indigenous land. So they can push it through without consent."

"Oh, Afa, this isn't fair. The tribes said they didn't want this! They said no!"

Afa shook her head and sobbed. "Why did they go to all this trouble of asking them in the first place? Now I believe that the most important thing throughout this whole process is that

Indigenous Peoples have agency over their land. To compensate for the centuries of human rights abuses, not having a say over their land, negligence from the international community when their forest was disappearing. And now we need them because we want rain. I want rain for them too, but now I feel that by advancing the rain program, I am working against them, and I don't want to do that. I'm on their side."

"I see it your way, too, Afa. That's why you must fight it. You have to."

"I will fight it. I will." Afa felt her resolve grow.

"And I will fight it with you. I will support you! I will support the tribes in any way I can. What can I do right now?"

"Give me the number of that lawyer, the one you told me about who worked with the Sarayacu to stop the big oil project in the 2010s. I'll start by talking to him."

"That's a great idea. I will text you his number. Tell him that you're a friend of mine and explain that you are seeking consent in the Amazon – explain your situation. He will know what to do. He has been fighting for Indigenous Peoples for 50 years."

"I will contact him when I get back to New York. Maybe I can go back to Ecuador and meet with him. Thank you, Vishnu. Thank you for this information."

"Of course. I will call Augustine right now and tell him to expect your call. I will do my best to explain the situation. After you talk to him, keep me in the loop. Let me know what he says. I believe in you Afa, and I believe in the fight. I believe you can do this!"

"Thank you for supporting me. I'll let you know what he says for sure."

"Do you want me to come to Brazil for a few days?" asked Vishnu.

"No. I'm leaving tomorrow. There's no reason to stay here. I think the Yanomami and Kayapo have already left Sao Paulo."

"Ok. Let's stay in touch, then, alright?" Vishnu paused. "I'm thinking of you."

Afa paused for what seemed like an eternity. "Vishnu. I need to ask you something," she finally said.

"Anything. Go ahead"

"What is your arrangement with your fiancée? You mentioned before that there was an understanding? I mean, you tell me that you are thinking of me, that you want to visit me. But I don't understand how it works in your relationship."

"Well, Afa, we have an open relationship. She lets me be with other women romantically while I am traveling, away from home. She knows I have other romantic partners outside our relationship. Because I spend so much time traveling, it seemed to make sense for both of us. I love my lady, but I also can have special connections with other women, of flesh and spirit, like you. Having a romantic connection with another woman doesn't change the way I feel about her. In fact, I love her even more because she lets me be with other women. I just know I am very attracted to you because you have a beautiful mind and body. And I want to be with you."

"Well…I want to be with you too. But as I hear you describe what sounds like…polyamory, and I know that's not as wild as it maybe once was considered, but for me, I just feel that I want to be the main piece, the only piece, not the side piece." As she said this, she heard Aliah in her head explaining how polyamory was the new normal. She felt the need to keep explaining her choice. "You see, in all my past relationships, I was never chosen. There was always something that stood in the way. I know for myself I want a man who chooses me. Just me. I know you and I are attracted to each other, but I've thought about it, and at the end of the day, I realized that you're not available to me in the way I want a relationship. This arrangement with you isn't what I want."

"I get it, Afa. I understand what you are saying. It's not for everyone. I just really care about you. I see you as special, and I can see us growing together," said Vishnu.

"That's the thing. I see you as special too. But for me, it can only be as friends and colleagues. I really value working with you."

"Yes, we are colleagues and friends. I guess I'll have to be happy with that," said Vishnu, sounding disappointed.

Afa realized having said that, she had let go of a big weight, and she sighed with the relief of it.

"Vishnu, I need to plan out my next steps. I've got to get off the phone."

"Afa, I just want you to know that you are my first thought when I wake up and my last thought before I fall asleep."

"Well, although I like hearing that, that is not how colleagues talk to each other." Afa had to be firm.

"Okay, okay, colleagues. I got it. Bye, Afa."

"Bye."

Afa was able to change the frequency of her mind very quickly. Removing the question mark about Vishnu proved to be very freeing, and leaving behind her complicated feelings about him, she set her mind to the task before her. She moved forward and thought of her possible next steps. Without talking to the lawyer for the Sarayacu, she wasn't quite sure what her next move was, but she knew that she was going to try to go see him, and she wanted to do it under the radar, without Kwame or anyone at the UN knowing. What she needed was to take some time away from the UN to help the Brazilian tribes, the Guyanese tribe, and she realized that she needed to ask Kwame for some time off. She needed to do it right away; it was urgent.

She picked up her phone and dialed Kwame. With only a one-hour time difference between Sao Paulo and New York, it was early afternoon for him. He picked up immediately.

"Hi Kwame, I just finished the meeting with the Kayapo and the Yanomami."

"Hey, Afa, how did it go?" He asked gently, which Afa thought was uncharacteristic of him, given the tone of their recent conversations.

"They don't want the program. They didn't even let me explain about the silver iodide being harmless," said Afa, matter of fact.

"I see. Well, we knew that was a possibility, based on your interactions with the Guyanese. The tribes are all organized and talk to one another." Kwame spoke with little emotion, his voice flat.

"I really wanted them to sign the agreement. I believe the rain program would do so much good," said Afa, reassuring her boss they were on the same side.

"Yes, of course," said Kwame. His voice was still like the dead sea, without any movement or waves.

Afa was taken aback by Kwame's unusual calmness. When was he going to tell her that the program was moving forward without consent? He was probably pleased with the Brazilian and Guyanese governments' decision to move forward without the consent of the tribes, thought Afa. As the largest Amazon country with a healthy economy, Brazil had a lot of power in the region. Brazil's actions could have possibly affected how the UN made their decision.

Afa considered what would happen next, but she knew she was not going to push to be understood by Kwame anymore.

Something had shifted inside Afa; the rain program was not her goal anymore. She would find her own way to support the decisions of the tribes to not have the rain program on their lands. She was at peace with them not giving their consent. She stood by their decision because it was theirs to make about their own lives, their own land, their own futures. She was going to advocate for them in any way she could. Perhaps that made her a double agent, but she knew, finally, where her heart was.

"This whole ordeal has been very hard on me, Kwame, as you know. I have failed to get consent, and I believe I need to take some time off." Afa figured if she used his language, "failure," he would go along with her reasoning.

"I understand. I was going to suggest the same thing, just for you to recuperate. Why don't you come back to New York where we can debrief about Brazil formally. We can discuss where things stand, and then you can take some time off."

"Sounds good. I'll fly home tomorrow, and I can come into the office the day after."

"How much time do you feel you need?"

"I don't know, maybe three months?" Afa was trying to calculate how much time she would need to travel to Ecuador and meet with the lawyer.

"Three months?" For the first time, there was some expression in his voice. "That's a long time. You would have to bring Nick up to speed so he can take on some of your responsibilities. I don't know how much vacation you have; would you take unpaid leave?"

He's probably pleased to be rid of me for three months, Afa thought. She didn't feel insecure about her position at the UN.

She had proven herself over the last 10 years, and it was very hard to fire someone at the UN. She knew she would have her job when she came back. "Yes, I have about a month of vacation and then I will take two months unpaid leave."

"Okay, sounds like a plan. You need the rest. I know this has been a lot."

He isn't going to tell me about Brazil till I get back to the office. Fine. I won't bring it up either, thought Afa. She doubted he had true compassion for how hard this had been for her, his words no longer meant anything to her.

The next day, in her hotel, she was garnering up the courage to call the lawyer Augustin Leon, a very important man in her eyes, and see if he could help the Brazilian tribes with their case. Brazil had so much power at the UN and the Guyanese were likely to fall in line with them. The Brazilians would be her focus. The Brazilian government had not even provided a chance for informed consent, they had made the decision before she even met with the tribes, overriding their legal rights in a way that was egregious. Vishnu had told her he could help.

The phone rang in an office building in the capital city of Quito in Ecuador, the personal cell phone of Augustin Leon.

"Hola?"

"Um, yes, hola. Hello, Mr. Augustin," Afa stumbled over her greeting. "This is Afa. I am a friend of Vishnu's; he mentioned to you that I would be calling you about my work at the UN?"

"Ah, yes, Afa. From the Islands. Working at the UN to get consent from the tribes for the rain program. Yes, I was expecting your call."

"Oh, good. Well, through my work at the UN, I have learned that the government of Brazil is going ahead with the rain program despite the fact that the Indigenous Peoples in Brazil were not given a chance to give their consent. And I am familiar with your work in the InterAmerican Court of Human Rights, and the other tribes that you have helped fight for consent. I know that is your specialty, your area of focus. And I was wondering if you would be willing to take on the case on behalf of the Brazilian tribes that did not give consent."

"Well, I guessed that is what you were going to ask me. And yes, I am willing to work on this case. As you know, the InterAmerican Court of Human Rights guarantees them FPIC – Free, Prior and Informed Consent. Brazil is a signatory to the agreements that lay the foundation to the court, so they are bound to the court if it rules against Brazil. I can reach out to the Yanomami and Kayapo. I have crossed paths with them before."

"Oh, that's great." Afa breathed out a sigh of relief. "I took off three months from the UN, and I was wondering if I could come to Quito and work with you? I'm not sure what I can do, what I can offer, but I want to do something. And I want to keep this private, without my colleagues at the UN knowing what I am doing," said Afa.

"Well, I have work you can do, but it would be different from your work at the UN. I would need someone to research all the case law in the Inter-American Court. I would need a research assistant to find legal precedent that we could use to argue that the Brazilian tribes have the right to consent, that it is a human right in the Americas, superseding Brazilian law. Everything is online, so you just need a computer, and all the case law is public information. I can craft the arguments; I would just need someone to go through all the case law of the last 50 years. And yes, I understand why you would want to keep your work with me private, under the circumstances," said Augustin.

"Yes! I'll do it! I'll come to Quito and research the case law. This is what I was hoping for – it's fantastic. Thank you. When can we get started?"

"Well, as you know, I work at the Center for Social and Human Rights in Quito, so I would have to go over my plans with my colleagues, and then I can contact the Yanomami and the Kayapo. Maybe we can get started in a few weeks? I am working on another case right now so I want to make sure that other people in the office can take over my current case. So nobody in the UN knows what you are doing?"

"No, they don't. The only person who knows is Vishnu, who connected me to you."

"I see. So," Augustin paused. "I am harboring a fugitive?"

Afa laughed. "No, not exactly." Then she thought for a moment. "Well, I guess in a way, yes. I just need to do the work that the UN was founded to do, but my supervisors have become so idealistic about the rain program, they don't see that they are overlooking the human rights of the people that the UN was designed to protect. I do feel like a double agent."

"Yes, the road to hell is paved with good intentions."

"Indeed!" said Afa.

"Do you know where you'll stay while you're here?"

"No, I don't know. I guess I can stay at a hotel. But that will get very expensive. I'm not sure..."

"Well, my brother Jorge has an in-law unit in the back of his courtyard, behind his house. He rents it out, and I can see if it is available for the next month or two?"

"Wow, Augustin. That's so kind, I'd really appreciate that. Thank you so much, I'm truly hopeful now. Gracias."

"You are most welcome Afa, I will see you in a few weeks. You can work out of my office in Quito. We have an extra desk that the research assistants use. They are usually college students, so it is nothing fancy. But you can do the work from here. I am sure there are things we will need to discuss in person. Just out of curiosity, which of the tribes you worked with said no and what were their reasons?"

"The Guyanese Macushi were concerned that the chemical that we shoot out of the rocket for the cloud seeing process, silver iodide, would poison them and hurt their spirits – more specifically, their Water Mamas – living in the rivers."

"Ah yes, I have worked with many indigenous tribes in my life and some tribes have preserved their spiritual beliefs, while others let those beliefs go and chose secular modernization. I can tell you with certainty that where there are spiritual beliefs adhered to, the forests are always better managed from an ecological point of view," said Augustin.

"That makes sense to me, I understand after hearing their folklore."

"And did they consult their shaman?"

"Yes, they did. He said the spirits would be hurt."

"Ah. So you cannot go against the shaman. You would not have been able to convince them that the rain program was a good idea. So, what about Brazil, what was their reason?"

"At first they mentioned not wanting to be poisoned, but I think it was really their distrust of the UN in general that prevailed as their major reason for saying no."

"Well, yes, that makes sense. The UN represents an alliance of global powerful countries that want something from them. They are skeptical of anyone wanting something from them at this point."

"Yes, I get that now. I was naive at first. I assumed I would just explain the program and they would be so happy to have rain again."

"I wish it was that simple. To be honest, I want the rain for them too. I wish they had signed on. But I am a human rights lawyer, and it is my job to defend their rights, which have been violated," said Augustin.

"Again." Afa sighed. "Okay, I will plan my trip to Ecuador, and I will be in touch with you as I finalize details."

"Sounds good! Give me your email address and I'll send you the information you'll need to book your flight. I look forward to working with you," said Augustine.

"Oh, I feel honored to work with you! Thank you so much!" Afa gave him her information and hung up. She hadn't felt this spirited and hopeful in a long time.

SEVEN

Ecuador was unique in that along with the remaining rainforest patch intact, it offered high altitude mountains and volcanoes, as well as a steep terrain that put on a stunning view for those coming in by plane to the capital city of Quito. The country was inhabited by Indigenous Peoples who had red, rosy faces, as the blood in their bodies had adapted to the enormously high altitude. She had read about the elevation in her research of the city, and she had found that many visitors frequently suffered altitude sickness. The remedy for the nausea was coca leaf tea – the same leaf from the plant that produces cocaine. Drinking the leaf in small amounts was not enough to create a cocaine high. It would take two garbage bags of coca leaves to equal one bump of coke. The white powder was highly manufactured and concentrated and did not resemble the coca leaf tea.

She was staying at a hotel for a few nights before staying at Augustin's brother in law's unit. Her hostel – The Americana – was in the American/ European neighborhood in Quito called Gringolandia. There were signs for "Dutch Bicycle Company: Visit the Volcanoes!" Afa thought that sounded good; maybe she would go to Cotopaxi, a volcano surrounded by Indigenous Peoples she had read about, not far from Quito.

After putting her things in her room, she walked to the hotel restaurant. The only person she had told what she was doing, besides Vishnu, was Aliah. Her plan was to help Augustin for a month or two, completely without anyone else knowing.

"Hola. Can I order one guanabana juice please?" She asked a man who looked like he might be indigenous. He was shorter and had narrow, kind eyes, with straight, jet black hair.

"Si, Senorita. Have you ever had guanabana before?"

"No, I have not."

"Do you want to see the fruit?"

"Yes!"

He brought out a green, callused, and bumpy skinned fruit cut in half, where the roundish fruit was sectioned off in white colored pods. He pulled apart one pod, and inside the flesh of each pod, there was a black seed. It was insane looking! She had never seen anything like it. A few minutes later, he presented her with a glass of the thick, white juice that had the frothy and thick consistency of a smashed banana. She wondered how they had extracted all of the big black seeds that she had seen in the pods. It was so fragrant, it smelled like a mix of pineapple and strawberries. With one taste, she knew she was going to have a glass of this juice every day that she was there. She quickly googled the fruit on her phone and saw that it was being studied for its medicinal properties as it contained chemicals that might fight cancer.

"Thank you so much for the juice. What is your name?" asked Afa.

"Pedro. It is nice to meet you. Are you here for vacation?"

Afa didn't know how to answer this question. She figured he wouldn't be talking to anyone who knew anyone at the UN. Initially she wanted to keep her presence on the down low. But he felt safe. She would take the risk and tell Pedro a little bit about her situation.

"No, I'm not on vacation. I'm here to work with the Center for Social and Human Rights here in Quito," said Afa.

"Oh, I have heard of them. They're a great organization! They've helped many Indigenous Peoples. In the 2000s, they helped the people of Sarayacu keep their lands out of the hands of the petroleum companies. I am Kichwa from Sarayacu. I go back and forth to the patch, but I work here in the American part of the city."

"That's wonderful. Yes, I have read about the Sarayacu. Mario Melo was the lawyer that fought Ecuador in court, claiming the state did not give the Saracuku consent to allow oil companies to start drilling. The court ruled that Ecuador had to stop all activities on their lands," said Afa.

"Yes, all the Sarayacu know Mario Melo, we were all sad when he passed. But I know that there are other lawyers that continue to work with the Sarayacu and other tribes in the Amazon. That's wonderful that you will be working with the organization. Welcome to Ecuador! It is wonderful to have you with us! I'm sorry, I didn't ask your name..."

"It's Afa, my name is Afa. I will definitely be back for more juice! Thank you!"

This exchange with Pedro felt like a good omen about her stay in Quito. The American neighborhood was close to Augustin's office, so Afa finished her juice and used the navigation on her phone to find the Center for Social and Human Rights. She walked through various crowded streets in pastel colors, cement and exposed brick for smaller stalls, and high-rises where the offices were located. Noticing one or two Chinese electronics stores, due to the large Chinese population in Quito, she stopped inside of one that looked like a small warehouse, but was messy like a garage sale, with boxes everywhere, some of them open. She perused until she found an alarm clock, a vestige of the past. She knew she could use her phone, but there was something charming about having an analog alarm clock while working abroad.

Continuing her walk, Afa passed a Swiss Hotel, and she couldn't resist poking her head into the bakery. The smell of confectioners' sugar and sweet pastry filled her nostrils, making her mouth water. The juxtaposition of this sweet smell to that of the fried onions and beef empanadas wafting through the streets called to her sweet and salty desires, and Afa wanted a little of each. She glanced around at the extravagant European pastries and cakes that called to mind delicate Chantilly lace, and she made a mental note to pop back in for a treat later since the hotel was only a few blocks from the office.

When she arrived at the three-story office building with shiny glass exteriors, she buzzed the third floor and was let in. As she walked up the stairs, she felt a familiar, nagging worry – would she be competent as a legal researcher for Augustin? Would she have anything to offer him? He was used to working with legal experts and researchers; she was a UN negotiator, representative, and liaison. Could she be successful at producing what he needed for the case to move forward? She responded to her doubts and hesitations in her own head. I got this. Fuck yeah. I've got what it takes. She squared her shoulders and entered his office.

When Augustin Leon walked into the room, the first thing Afa noticed was a certain energy, a presence of someone who was very evolved in his sense of self. Dare she say enlightened. It was as if he vibrated on a higher frequency. Afa was slightly intimidated.

"Afa! Nice to meet you!" Augustin extended a hand, a broad smile that reached all the way to his kind eyes taking over his face. "How is your Spanish?"

Afa took Augustin's hand and shook her head, "Not good, I'm afraid…"

"Ah, well, not to worry!" Augustin assured her with a wave of his hand and a warm chuckle. "All of the Inter-American Court and Commission Cases are published in English and Spanish. You can do your case research in English and then find the corresponding sections of the cases in Spanish when you present your findings to me."

"Ok, no problemo," Afa smiled. Her attempt at a joke making her feel foolish before she'd even finished the "o" – but Augustin gave her a wink and a nod, laughing a little. Beads of nervous perspiration formed on the nape of her neck; she hoped her deodorant wouldn't fail. She wondered if her blue eyeliner was dripping down her eyelids.

"This will be your desk and your computer. I think it will work for you for a few months, if that's how long you are planning to stay with us."

"Yes, that sounds right."

"And the UN doesn't know that you are here? So we can't mention you in any publications or give you credit for your research. Are you fine with that?"

"Yes, of course. I'm not doing this for credit. I am doing this for the tribes that trusted me with their right to consent which I said I would protect."

"That is very admirable. We are happy to have you here with us. Sit down. Let me start at the beginning and give you some background on the work you are doing. All of your case work will be about finding case precedent for Free, Prior, and Informed Consent, or FPIC, a right that has been set up by the Inter-American Court of human rights for Indigenous Peoples who are subject to government or corporate large scale projects on their land, such as mining, hydroelectric dams, petroleum extraction, or now, the rain program. The majority of Latin American and the U.S. recognize FPIC as a part of the Organization of American States. The very first case in the Inter-American Court that set up FPIC was the 2001 case of the Mayagna Community of Awas Tingni v. Nicaragua. The Awas Tingni decision by the Inter-American Court of Human Rights was the first occasion on which an international court with legally binding authority found a government in violation of the collective land rights of an Indigenous people. The Inter-American Court concluded that Nicaragua had violated the rights of the Mayagna community of Awas Tingni by granting a logging concession within the community's traditional territory without obtaining its consent and after ignoring the consistent complaints and requests of Awas Tingni. This case is the first of its kind – it was the landmark decision that all our work is based on – securing that Indigenous Peoples have consent that is backed up by international courts. That is what we call precedent. These cases set up precedent for future cases. Are you following so far?"

"Yes, I am. That was the first big case for FPIC," said Afa with the eagerness of an A+ student.

"Ok. Good," he said while nodding, giving her a look of approval. "The case of Sarayacu v. Ecuador was equally important. The Sarayacu case was argued by my mentor Mario Melo. He was the lawyer at the Center for Economic and Social Rights, the organization that was the predecessor of our organization. In 2012, the court ruled in favor of the Sarayaku in the case of Sarayaku v. Ecuador, affirming and upholding the right of free, prior and informed consent of Indigenous Peoples. The ruling took a decade-long legal battle as the Sarayaku fought the government for allowing a foreign oil company to encroach on their traditional lands in the mid 2000s. By ruling for the Sarayaku, the Inter-American Court agreed that the state had imposed an oil project without previous consultation or consent. These two are the most important cases for FPIC. There are many other cases that mention FPIC, and I will need you to look through all the case decisions and find precedent and pull the language that we need, based on the facts of the current case. You are well poised to do this research because you know the situation in Brazil. From what Vishnu told me, the Macushi in Guyana also did not give consent to the rain program, but Brazil will be our main target. The government of Brazil did not even allow for the FPIC rights of the indigenous tribes in Brazil, so we have a strong case in Brazil."

"I see," said Afa, trying her hardest to take all the information in. Her eyes had started to squint.

"So how are you doing so far? Do you have questions? Do you need a break? I'm throwing a lot at you."

"I'm following, but maybe I can step out and get a quick bite to eat and be back. Some sugar would help me. Can I get you something from the Swiss Hotel bakery?"

"Ahhh, she is a golosa!" said a woman who was seated nearby and inserted herself having overheard the conversation.

"Ah, this is Marta. She is our office assistant," said Augustin, as he nodded in Marta's direction.

"What is golosa?" asked Afa.

"Someone who loves to eat, like a glutton!" Augustin laughed. "Don't get full, we are going to take you out for ceviche for your first day here. The ceviche place around the corner is the best in Quito."

"Okay, I will be right back..."

Afa felt a bit overwhelmed. When she arrived at the Swiss Hotel bakery, which she really hadn't been able to stop thinking about despite the overload of information, she could not believe all the custard and cream that filled all the flaky pastries and dense cakes. She bought a decadent piece of white cake with almond cream frosting and sat down and ate the whole thing. It was as if the sugar and cream was giving her the energy she needed to get started. She was hyping herself, trying so hard to not let any self-doubt permeate her thoughts. Who knew the Brazilian tribes' issues with the rain program better than her? She had something to offer. Yes, she did! She walked back, readying her brain for taking in as much knowledge as she could fit.

When Afa returned to the office, Augustin was speaking to Marta, who had called her "golosa."

"Afa, please join us, you would benefit from hearing this conversation," said Augustin. Afa nodded and gingerly took a seat at the table.

"I am just handing off my current case to my colleague so that I can work on the Brazil FPIC case. Before we sue Brazil on behalf of the Yanomami and Kayapo, we have to transfer our current work to other lawyers who have experience. Señora Afa, do you want to hear about our last case that we are handing off?"

"Yes, of course," Afa said.

"We have a case here in Ecuador where Yasuni National Park was opened to petroleum extraction, a decision made by the Huaorani indigenous people who lived in the park 30 years ago. After several oil leaks from faulty pipelines, water pollution and cancer cases, as well as petroleum chemicals leaking into the water causing infertility and animals dying, the Huaorani

Indigenous Peoples sued Conoco, the oil company that was leading the extraction. The Haorani decided to settle the case for a large sum of money, with the condition that there would be a gag order forbidding any member of the tribe from discussing the details of the case. No one knows the details of the settlement besides the Indigenous Peoples, Conoco, and us, the lawyers for the Huaorani. What makes this case so interesting is that the gag order includes the children, as young as five years old. They are not allowed to talk about the case for the rest of their lives. The court got away with this because there was legal precedent for a gag order for children stemming from an American case involving fracking. Are you familiar with fracking, Afa?"

"Yes, it's when you extract petroleum out of shale rock, but it leaves behind a lot of chemicals – it is a chemically intensive process."

"Yes, exactly. In one of the most controversial legal cases around fracking in the whole world, in 2011, it was ruled that two children in Pennsylvania would live their lives under a gag order in exchange for a settlement of $750,000. The settlement was between the Hallowich family and Range Resources Corp. – a leading oil and gas driller. The gag order barred their son and daughter, ages ten and seven at the time, from talking about fracking or where it took place, for the rest of their lives. Gag orders on adults are typical in settlements reached between oil and gas operators and residents in Pennsylvania, but the company lawyer's insistence on extending the lifetime gag order to the Hallowichs' children was extremely rare at the time. In their legal case, the Hallowich family accused the oil and gas company of destroying their ten-acre farm in Mount Pleasant, Pennsylvania and putting their children's health in danger. The Hallowich children lived next door to major fracking operations and the family was exposed to four gas wells and a wastewater pond, which contaminated their water supply and caused burning eyes, sore throats, and headaches, as well as killed all their animals. Through settlements and gag orders, companies bury the evidence of contamination and public health problems associated with fracking. So 30 years ago, when Yasuni Park was still a standing rainforest, before it dried out, the Indigenous Peoples agreed to that gag order, and we have been trying to fight it. But I am handing off that case in order to work with you."

"That sounds like a very important case as well, and I am grateful that you are taking on the Brazil case. Thank you."

"We will sue in the Inter-American Court of Human Rights. I will talk to the Kayapo and Yanomami, who will be the plaintiffs in the case. Since they both legally have demarcation rights to and live in Brazil, either one or both could be a plaintiff to represent their right to consent. Do you believe you have a good relationship with them to discuss which one should be the plaintiff?"

"In my last meeting with them both, they were very angry at me. I don't believe that they would want to talk to me again since they see me as representing the UN."

"No worries. Maybe while you are here, you can visit the Kichwa from Sarayacu again and get support and learn how to communicate with the Brazilian tribes. I visit them by helicopter a few times a year to see how they are doing. They are inspiring, a true success story."

"Yes, I loved staying on their patch, I drank chicha with them. When I felt ill, they used snake venom on my stomach and had me drink something red that was the sap of a tree."

"Ha, yes, Sangre de Drago. I have also drunk that when I felt sick while staying with them," said Augustin.

"Did you get to spend a lot of time with Mario Melo? I am sure he was such an inspiring lawyer for the Sarayacu. I bet they really appreciated him," said Afa.

"Oh yes, he spent a lot of time with them. They asked him to do an Ayahuasca ceremony with them and to gain their trust, he did. He told me that he spent the first three days vomiting, and then the ceremony began. He was guided by a shaman and said he had such visions and realizations and healing. He left the Amazon to come back to Quito after a week and he told me that his wife said that his eyes were different for three months! Mario admitted to me that he felt the effects of the ceremony for months after ingesting the Ayahuasca. When I shadowed him and went to visit the Sarayacu, they offered me the Ayahuasca ceremony, and Mario forbade me from taking part because he said he needed me to be alert and that he could not afford for me to be under the influence of the Ayahuasca for three months! And since he was my boss, I had to listen," said Augustin with a wistful look in his eyes, as if he was imagining what could have been.

"Wow, that is such a great story," said Afa.

"Yes, Mario had a beautiful, kind wife and two daughters and he sacrificed a lot. He did not see his family very often as he was always in the Amazon, what is now the Pastaza patch, for weeks, months at a time. He really gave a lot of himself. He was an amazing man. I have empathy for Mario because I also can spend a lot of time away from my family. Do you have a family, Afa?"

"No, I don't. I'm not married, and I don't have children. I'm not sure when or if that will happen for me, but I'm not in a rush. Right now I just have a mother and a father. We are immigrants. We left the island of Tuvalu before it went underwater from sea level rise, before the country went extinct."

"Interesting. How do you think your past relates to the work you are doing now? Is that what motivates you? Losing your home?"

"Yes, losing my home, my country. But more than that, I relate to the fear of the unknown, not knowing what will happen if you take one direction, or another direction. What is the right way to go? I empathize with the Indigenous Peoples who are losing their land due to the disappearing Amazon, the lack of rain. But I also empathize with their fears of the rain program, their resistance to the program. When my family was on the island, the government began to erect sea walls all along the shores to prevent sea level rise, and that gave hope to many people. Many climate scientists were saying that sea walls were a great method of adaptation. But the water seeped into the ground and traveled through the earth and started to erode the pipes. There was saltwater in everything; all the crops were ruined, and so the sea walls didn't work. But people were very hopeful that they would, so they stayed longer on the island only to face disappointment and hardship as it was harder to leave the island the longer they waited."

"I see. What a tragedy, to lose your home in such a way. I'm so sorry. This is all very interesting to know about you Afa." Augustine studied her for a moment, and she recognized compassion in his eyes. He shook his head and tapped the globe on his desk as if to say, what have we done, climate change. "Well, moving on to your time here. You will enjoy staying in my brother Jorge's in-law unit, with his wife Lucia and their daughter Omara, who is only four years old. They have a housekeeper, an indigenous woman who takes care of the house, cleans, cooks, does laundry, so there will be someone to clean your room and help take care of you. She comes with her daughter, who plays in the yard. You will be responsible for your own meals, but you can use Jorge and Lucia's kitchen, which is across the garden patio in the main house. I have spoken to Jorge, and he says that you can arrive today after work to meet them, to stay in the in-law unit."

"I was going to stay at the hotel for a few more nights, but I can go to Jorge's house tonight. It will save me some money," said Afa.

"Good. Now, let's get you situated with the court website so you can read all the cases involving Indigenous Peoples in similar situations. We want to find rulings that support our argument that the Yanomami or Kayapo have FPIC over their land, despite Brazil's claim that they are overriding their rights in favor of national security. They are violating international human rights law by arguing that the rain program is a matter of national security," he said with a grave expression.

"Ok, great. Should I get started now?"

"No, let's log on, but first we are going to go have the best ceviche in Quito! Put down your things and let's go!"

As Augustin and Marta walked through the streets with Afa trailing behind, she noticed many tall cement office buildings mixed with quaint colorful storefronts with tin roofs, and corner stores that seemed to be filled to the brim with soda, alcohol, toilet paper, and crackers. The fried onion smell filled her senses again, the smell of empanadas being cooked and sold by an indigenous woman whose storefront was near the office. As they passed her, she said something in Spanish that Afa did not understand.

"What did she say?" Afa asked Augustin.

"She said she likes your shoes."

Afa looked down. She was wearing white sneakers with pink and camel color embellishments. She could see the empanada seller's shoes which were very worn and had holes in them. She wondered what her size was and if she should give her the shoes before she left but didn't know if that would be insulting. She would ask Augustin if that was appropriate to do.

As they reached the ceviche restaurant which looked like a wood hut painted in blue, she saw people sitting outside on wood communal benches with bowls of ceviche with a side of popcorn.

"I suggest you get the typical ceviche, and if this is okay, I will order it for you," said Augustin. The line went very fast and soon they were sitting down at the table with the colorful metal bowls in front of them, filled with white fish, red onions, and lime juice. Afa took a spoonful to her mouth, and it tasted amazing. Fragrant citrus mixed with the fleshy ocean tasting fish.

"This tastes so good!" said Afa.

"Ah, yes. Marta and I come here once a week. You can join us if you like!"

"Yes, please! I would love that," said Afa enthusiastically, digging back into her ceviche with relish.

***

Jorge's house was indeed on a steep hill and painted red in the Dorado neighborhood where the streets were made of cobblestone. It appeared to be an older neighborhood. Afa knocked and after what seemed like a minute, a beautiful woman with curly hair, olive skin and a red poncho opened the door.

"Hola, I am Lucia, Jorge's wife. We have been expecting you. Please come in, I was just watering the plants in the courtyard. So many plants!" She had a frenetic energy to her; the type of person that did a lot of things at the same time, giving off a beautiful chaos.

"Thank you so much. It's such a pleasure to meet you."

"We thought that you could eat dinner with us tonight, since it is your first night. We are eating crabs, and we were hoping that since tomorrow is Saturday, we could drive you around the old historical part of Quito where all the old buildings and churches are, for some sightseeing."

"I would love that, thank you!"

"Jorge!" she called out to her husband. Afa guessed that the longer string of Spanish words she called out to him was her telling him to come to join them in the foyer. Jorge appeared, a man of medium stature, with a swagger in his step. He had a beard and was dressed in a disheveled cream button down shirt and camel-colored trousers, like something he wore to the office that he had yet to change out of. He had a warm and gentle expression.

"Well, well, it is not every day that we have someone from the UN staying with us! Welcome! We can talk about your work here in Quito tonight but first let's show you your room so you can put your things down and see where you will be staying for the next month."

Afa followed Jorge and Lucia through the house, across the cement floored courtyard adorned with an array of wild looking tropical flowers that she had not seen before. They arrived at a tiny cottage at the end of the yard with a wood door that didn't quite fit the frame.

"The bathroom is outside the unit," said Jorge. Afa turned to see a tall, narrow building made of wood on the side of the courtyard. "You will take your showers there, and there is usually hot water." Afa thought this would be fine, as she had used outdoor showers on the island. "Now, please put down your things, rest, and dinner will be in about an hour. Usually, you will have your meals on your own, but it is your first night and you are helping my brother Augustin. He says you brought him the newest case! You are a special visitor. Any questions before we leave you?"

"Um... no, no... I think that's it. Thank you so much for everything, I'm excited to have dinner..."

"Mama! Mama!" shouted a small child.

"Omara!" called Lucia. She spoke in Spanish in a soothing voice, and it seemed as though she called for the small child to join them in the courtyard.

"Now you have met everyone. We will leave you alone to rest. Welcome!" said Jorge as he left the unit and closed the awkwardly misshapen door.

"Thank you so much Jorge," said Afa as she bowed her head. She waited for him to leave the room and then she threw herself on the twin sized bed. She felt peaceful; she was doing the right thing.

***

A few weeks later, on one of her weekends off, Afa was eating watermelon she had bought out of the back of a truck on a dirt road outside a butterfly preserve in the town of Mindo. She was a few hours from Quito by bus. She was surprised when her cell phone rang – Augustin would not call her on a weekend. She looked down at her phone and saw that her colleague Nick was calling. She was puzzled.

"Hello, Nick, how are you?"

"Hey, Afa."

"I'm on vacation from the UN right now," Afa started to say. She hadn't told Nick anything about her work in Quito and she was slightly paranoid that he would find out.

"I know, and I'm sorry to interrupt, but I thought you would want to know that the New York Times is running a story about the Brazilian and Guyanese tribes attempting to prevent the rain program from being adopted throughout the Amazon. The article mentions that the Brazilian tribes have sought legal representation. They called here for comment and Kwame gave the comment. I thought you would want to know, to hear it from me. I don't like how Kwame keeps you out of the loop. His comment was very pro-weather modification and anti-indigenous."

"What? How could he make the comment without consulting me? The tribal agreement is my work, not his! Why didn't someone reach out to me?"

"Because you are on vacation, ostensibly, Kwame said that he would be making the comment. He took this opportunity to make the tribes look anti-climate change modification."

"God, this is awful. Such bad optics on the part of the UN. When is the story going to run?"

"In the next few days."

"Okay...Thanks so much for letting me know."

"Are you having a good vacation? Where are you?"

"I'm in Ecuador, and it's amazing. I'll be back in New York in a few weeks." Afa hoped he would stop asking questions. She liked Nick, and under normal circumstances, she wouldn't mind continuing the conversation, but she couldn't risk it.

"Well, I look forward to seeing you when you get back, let's definitely get together."

"Yeah. Okay, let's for sure do that. And I really appreciate you giving me the head's up."

Afa put her phone away and resumed eating her watermelon, thinking about how kind Nick was to always think of how these big decisions they were keeping from her affected her. He was considerate of her and seemed to care about her, and that felt good. She wouldn't mind getting to know him more.

As for next steps, she would have to wait for the article to come out to see exactly what Kwame had said. She went back into the butterfly preserve, and one of the guides called her over. He then smashed an overripe banana and placed it on Afa's hand. A small, brilliantly blue butterfly, one of the few remaining in the entire population, landed on her. Afa was briefly distracted by the beauty of the intricate and geometric mirror image patterns of both wings. So delicate and lacey. But this news about Kwame's decision was upsetting and changed the tone of what had been a very peaceful day.

In fact, since arriving in Ecuador, Afa had been having very peaceful weekends. Either she would stay in Quito, buy earthy looking vegetables from the market around the corner and cook something for herself, or she would go out of town for short trips. Many of the towns outside of Quito had schools with a big Nestlé sign on them. When she had asked Jorge why all the schools said Nestlé on them, he said that Nestlé paid to build schools in exchange for pumping out all the groundwater in those villages for their factories. Afa wondered if the school children ate Nestlé chocolate and chocolate milk every day.

Quito was an urban city, but it was surrounded by volcanoes. One weekend, she had gone up to the top of a small volcano near Cotopaxi with a Dutch bike tour company operating out of Quito. At the top of the volcano was a crater, with white plumes of steam coming up from a cone in the earth. They unloaded the bikes, and all the tourists biked all the way down the volcano. She had really been experiencing the local area, having gone to the Andes and walked with indigenous guides that were much shorter than her, with rosy cheeks from the thin air. She had gone to markets in the town of Otavalo and bought incredibly colorful, yet scratchy, wool blankets and shawls.

Now this news had reminded Afa that Kwame did not understand her work, and he had the power to shape the narrative. Perhaps Augustin would have some advice when she told him about the release of the article. Afa wanted to speak to him now but would wait till Monday morning. She didn't want to interrupt his time with his family on the weekend.

On Monday, Afa walked to work earlier than usual and checked the New York Times on her computer. The story was there, on the front page. Indigenous Tribes Attempt to Block Global Rain Project. She did a speed read of the article, scanning for Kwame's quote. "The Indigenous Peoples don't understand what is in their best interest with regard to the global rain program." It was like she was punched in the stomach. She felt nauseous. All her work at the UN had suddenly been reduced to this statement.

She had been completely dismissed and disregarded. Kwame had overlooked her extensive notes and reports detailing the nuances of their perspective, the complexity of their opinions, negating her work gathering information about the Macushi world perspective. She was gutted. But what did she expect? He was a technician, not sensitive to beliefs that were outside of science. And the tribes would feel that she had betrayed them. She felt guilty for not being able to shape the narrative from the perspective of the tribes. It felt like an injustice.

She kept reading and found that the journalist was somewhat empathetic towards the indigenous communities, asking, "Should historically oppressed people who have been beaten down for centuries, have the power to decide what is best for the planet?" Almost verbatim of what she and Aliah had discussed before, the question she had been asking since her quest began.

About 30 minutes later, while she was still waiting in the empty office for Augustin to arrive, she heard her phone ring.

"Hello, Vishnu!"

"Hi Afa. I saw the article and had to call you."

"Yeah, it's beyond horrible. I'm very upset. Has Marcia seen it? Noel?"

"Yes, Marcia has seen it," said Vishnu.

"And what was her reaction? Does she think I'm behind the quote? Do they feel I've betrayed them? Do they know I wasn't involved in the article?"

"Marcia was confused. She said to me, 'I thought Ms. Afa was a friend of the Macushi.' But I explained to her that the person commenting was your boss, and that you were on vacation in Ecuador and didn't know what was going on at your office."

"Vishnu, please tell her that my boss went against the work I presented to him, and that the opinion was not mine. I was not part of that quote. My boss and I are in conflict about this situation. I honor the Macushi perspective, and I conveyed that to the UN."

"I will explain to her that you and your boss do not agree, and that he went against your wishes to say what he said. It has been pretty upsetting for the community members that have all been in the discussions about the rain program with you."

"Yes, of course. I understand why it would be upsetting to them. But please explain! It's upsetting to me as well."

"I'm about to see Marcia for a meeting, so I'll talk to her."

"Yes! And talk to Noel too! Tell her I was not behind the quote!"

"Don't worry, I will," soothed Vishnu.

"Thanks, Vishnu."

After another 30 minutes of torment and empty office sounds, Augustin finally arrived. As soon as he walked in, Afa pulled him to her desk and showed him the article. He sat in silence and read through, scrolling the black words on the white screen with images of rain and maps of the Amazon and some photos of Indigenous Peoples with markings on their faces. To Afa, it seemed meant to conjure the image of a primitive, uncivilized group of people, those out of touch with modern times, primitive. It made her angrier.

"My, oh my. This is your boss? The one who is quoted here?"

"Yes, that's him."

"Well, I have good news. Things have moved quickly, and the Inter-American Court has granted us an injunction, which is temporary. It means that according to the international court, the state of Brazil cannot move forward with the rain program until the court has made a decision on the case. This is a win for the Yanomami and the Kayapo, for now. That doesn't mean we have a final decision; it is just a kind of urgent order to "halt" the process until the court reviews the case. There will still be more deliberations. It is the first step of a long struggle, and we must see if Brazil will act in accordance with the court's decision, since the court has no way to enforce the decision. Brazil has signed on to the American Declaration of International human rights, but that doesn't mean they will follow the courts' recommendations. But for now, we won the first battle. It's the first step, and it's a big one. Thank you for all your help and your service towards this case."

"That is wonderful news!" said Afa with a sigh of relief. "I guess it's time for me to go back to New York, at least for now."

"Yes, I think so. But we will keep in touch as the case moves forward."

"Thank you Augustin for giving me the opportunity to work on the case. It felt really great to be here."

"Of course. You have been very valuable to us. I will keep your work with us very private. I know you have had profound experiences with the tribes, and you care deeply," said Augustin as he clasped his hands and raised them to his chest.

"I do care. For me, it's about preserving their view of the ecosystem, not just preserving their ecosystem. But that is not how I started the program. At the beginning, I thought my job was to show them that the program was essential, necessary. But I realized working with the Macushi in particular, their self-determination, the cultural reasons that led to them not giving consent became the most important to me. I wanted to respect their perspective, their perception of reality, no matter what it was. And as for my boss, he didn't want to hear what they had to say. The people who have been so screwed over by humanity, finally have a say about their own humanity. They can finally have control over decisions that affect their lives, and I want to protect that, no matter how unpopular."

"I know. That is why you are here with us now," said Augustin, slowly and deep from his chest. "Thank you."

EIGHT

Afa was glad to be back in New York. She had missed the red brick buildings, the omnipresent scaffolding which just seemed to move from block to block, and her morning bagel; egg-and-cheese from the bodega on the corner which was open 24 hours a day. It was spring in New York, not hot enough to smell the rotting garbage on the street, not cold enough to wear a jacket. Afa would miss the jungle, the smell of damp plant flesh, leaves, stalks, and roots. While most of the world was in a constant drought, the remaining patches were still a bewildering rain-soaked jungle. Something you would have to see to believe. She would love to go back every now and then, but she was also happy to be back home in the city where the density, energy, and fast pace felt like home.

Afa was on the phone with Aliah. "Augustin filed for an injunction and the court granted one. That means that Brazil has been temporarily ordered not to move forward with the rain project until the court decides. But that is just the first step of the battle. Brazil might not respect the ruling of the Inter-American Court of Human Rights. And now I have come back to reality here in New York. Kwame wants a meeting with me. And not just a check-in. A formal meeting. There will be another person there, his name is Ramin. I don't know what the check-in is about, but I am paranoid that they know what I was doing in Ecuador, working against the rain project. I don't think there's any way he could know. I cut all ties while I was gone, and everyone thinks I was on vacation. You and Vishnu are the only ones who knew what I was working on."

"I'm sure it is fine. Maybe he just wants to talk to you about the next steps of the program."

"Probably. I don't like working with Kwame. I know this is just a job in many peoples' eyes, but our values don't align. He has pit climate change remediation against indigenous human rights. I just think there could be another way to do it. So we'll see what he has to say."

***

Afa walked into the familiar small conference room with tall glass walls in the UN building, overlooking the river. There were three tall glasses on the table with a pitcher of water. The room was cold and chilly. It seemed like a formal meeting, and even Afa's hands were sweating. After about 10 long, arduous minutes of waiting, Kwame and Ramin walked in. Ramin had white hair and deep-set eyes, thick eyebrows and olive skin.

"Hello, Afa, thanks for meeting with us today. This is Ramin Kalantari, and he works in the Department of Economic and Social Affairs here at the UN, an office of the UNDP. We called you here to discuss next steps."

"Hi Afa, nice to meet you," said Ramin.

Suddenly, Afa felt sick. Was he about to let her go, after all these years at the UN? Could he fire her? Surely, he couldn't, after all she had done for the UN. Did he know about her work in Ecuador? Cool sweat beaded along the nape of her neck and her heart felt like it was pounding out of her chest.

"As you are aware, the multilateral rain program may no longer work if the now high-profile international court case is ruled in favor of the Indigenous Peoples. The UN cannot violate international law, and we cannot rehabilitate the entire Amazon as one contiguous ecosystem without Brazil and Guyana. There is a high possibility that we are shutting down the Amazon-wide program as it is now, intended to bring rain to the entire former Amazon ecosystem. For now, the rain program will work country by country, on a small scale. This might not even work to bring the Amazon back, but it's our only shot. Every government that has consent from Indigenous Peoples will conduct their own rain programs as they see fit in the South American continent. Thus, your work on getting signatories to the rain agreement is finished."

Afa knew it. He knows about what I did in Ecuador! she thought. All the air left her lungs, and her muscles stiffened. Blinking rapidly, she took a sip of water, stalling. Her mouth was so dry, but she didn't know what to say. Kwame continued without a pause, seemingly oblivious to her state, which was a relief to her.

"I know you care very deeply about the indigenous right to consent, their perspectives, and their issues, so I contacted Ramin who works in the same office that drafted the United Nations Declaration on the Rights of Indigenous Peoples. I'll let Ramin take it from here."

Afa waited quietly, her curiosity now piqued.

"Afa, as a response to our inability to get consent from all the tribes, and the dire situation in the Amazon with the patches shrinking, we are going to abandon the current program and start a new one. But we will go about it a different way. We are envisioning a program where indigenous cultures and views can be integrated into ecosystem management. We are hoping that we can get their buy-in as co-leaders of the rain programs. The UN will strongly consider both the role that culture and science play in resource management. Indigenous communities will work very closely with the UN on a plan to restore the Amazon. Perhaps if they work alongside the scientists, and they are a part of the planning, they will be more likely to give consent. We believe consent is still possible and that is what we are aiming for. We have not given up, and our department will spearhead the effort moving forward. We will start a working group, and the first task of the working group will be to document current indigenous cultural ways of ecosystem management in the Amazon, focusing on the tribes that have most successfully and sustainably preserved their land. We will also document their resistance to the program based on their spiritual and cultural views that I have read about in the notes Kwame shared with me. This will all happen, obviously, with the involvement of the tribes. You would be working very closely with them. Perhaps, if they work directly and alongside reputable scientists that build relationships with them, we will have a different result this time. We see you as an asset to our working group, as you have on the ground experience with the tribes and their attitudes towards climate modification." Ramin paused, a warm expression in his eyes as he regarded Afa. "Does that sound like something you would be interested in? Would you be willing to work with our department?"

"Basically, this would be a reassignment, Afa, in an office that suits you better than climate modification," said Kwame.

Afa felt her tight chest melting as relief flooded through her. This was an amazing proposition. Her nervous sweat turned into excited perspiration. She could continue to work with all the tribes, but on their terms. Her excitement bubbled to the surface at the thought of traveling back to the patches, but this time, as an ally of the people she had grown to care about. Her response was ebullient and without hesitation.

"I would absolutely be interested! Thank you so much for this opportunity! I would love to join you in your department." Afa felt so confident that this assignment was right for her, she could hardly contain herself. Her voice resonated from deep inside her belly. Finally, she felt herself step fully into her place, as she knew that this reassignment was right for her.

"I'm glad to hear that," said Kwame. "You would start right away." Then he paused, looked out the window, before he turned his gaze back to Afa. "And I want to tell you something. I know we have had our differences working together, but I want you to know that I respect you immensely. I see your commitment and passion, and I admire it. I hope we can keep in touch." His voice sounded soft at the end, almost as if he was asking a question.

"Thank you, Kwame. I appreciate you saying so. And, yes, I would like that." She couldn't reciprocate the compliment, at least not yet. But she considered that Kwame had seen the work she had done, and he had reassigned her to a position that was right for not just her, but for the communities she had represented. That counted for something. Perhaps there was more to him than she thought.

"That's what I had hoped," said Ramin. "Here is some paperwork for you to sign, and I will send you more information about the department and the work you will be doing. That will go to your new email, which admin will set up immediately. I'm excited for you to work with us, Afa! I think you will bring a lot of knowledge and experience to the team! You will hear from us soon."

"I'm looking forward to it! More than I can tell you." said Afa. She meant every word.

The two men stood up and walked out of the room and Afa followed, feeling electric inside. She took one long look at the small conference room where she had so many meetings over the years. This was the end of an era. She took a sip of water from the water glass, tasting all the chlorine in the treated water, so different from the rainwater she had drank in the Amazon patches. She was so glad she knew the difference. She slowly put the glass down and walked out of the room.

AUTHOR'S NOTE

The State of the Amazon and Indigenous People's Human Rights

The South American Amazon accounts for over half the planet's rainforests, spanning nine countries covering an area approximately 25 times the size of England with approximately 65% of the rainforest located in Brazil. The Amazon holds 10% of the world's species biodiversity and 15% of its freshwater in river systems that drain into the ocean. The Amazon rainforest is often referred to as "the lungs of the earth" because it is responsible for absorbing large amounts of carbon (referred to as a carbon sink) from the earth's atmosphere and for producing a crucial amount of oxygen for the planet.

It is the presence of excess carbon in the atmosphere that drives the greenhouse effect and warms the planet, referred to as global climate change. In addition to being of great value to the planet, millions of Indigenous Peoples speaking hundreds of different languages inhabit the rainforest. Many of the Indigenous cultures possess intimate knowledge of the complex forest systems and medicinal plants which have yet to be documented by western scientists.

The Amazon is crucial to regulating the climate of the earth, yet it is being deforested at an alarming rate through a variety of means, and every year, deforestation increases. In August 2019 alone, 1,114.8 sq km of forest was cut down – the same area as Hong Kong. This does not include the large fires that were happening contemporaneously, mostly in Brazil where the government was unresponsive to the disaster at the time. Data has shown that the Amazon is not absorbing as much carbon dioxide as it used to because of climate-change-related droughts. In the major Amazon droughts of the last 25 years, on-the-ground monitoring showed that the forests stopped absorbing carbon dioxide from the atmosphere and dead trees released carbon dioxide into the atmosphere. At this current rate of degradation, scientists warn that the Amazon rainforest is shifting from being a carbon sink to a carbon source in the atmosphere, further exacerbating climate change and irrevocably breaking the water cycle.

Forests can recover from drought, fire, deforestation, and exploitation if there is adequate time, but we are experiencing mass removal, droughts, and fires – and we are not giving the forests time to recover from the disturbances. Scientists such as Thomas E. Lovejoy predicted that once a rainforest ecosystem is substantially removed through deforestation, the rest of the ecosystem will be unable to support itself and will start to die off on its own. This is referred to as forest dieback, a theory rooted in the work of Brazilian physicist Eneas Salati, who was able to prove in the 1970s that the removal of vegetation contributes to climate change. Scientists and researchers estimate that the tipping point for dieback in the Amazon begins once total deforested area exceeds 40%. Once the forest has scaled back past this threshold, it would be impossible for the forests to grow back. It is estimated that 18-20% of the Amazon rainforest has already disappeared.

Indigenous Peoples, also known as native, aboriginal, or first peoples, are groups of people who are the original ancestral inhabitants of a given region. The term "original" contrasts with groups that have occupied or colonized the area more recently. The United Nations defines Indigenous Peoples as inheritors and practitioners of cultures and ways of relating to the environment. Indigenous societies are found on every continent of the world.

Despite their cultural differences, Indigenous Peoples all over the world share common problems related to the protection of their traditional lands. In most cases, they are dispossessed of their use of traditional lands by private corporations and government. They fight for their systems of communal property to be recognized in countries where only private property is recognized. Often, Indigenous Peoples have alternative ideas to ecosystem management than the conventional ideas of the dominant society. Despite having their own systems of stewardship, they are not always in the power position to enact their culture-based systems of management.

Because Indigenous Peoples face threats to their way of life, the United Nations has created a large body of human rights laws and established courts and mechanisms to promote and protect the human rights of Indigenous Peoples. In 2007, The United Nations Declaration on the Rights of Indigenous Peoples (UNDRIP) was adopted by a majority of 144 states in favor, with four votes against it (Australia, Canada, New Zealand, and the United States). The declaration protects the cultural, economic, and political rights to resources on which indigenous cultures depend. The declaration is important because it guarantees Indigenous Peoples' human rights that are meant to override countries that fail to protect or violate the self-determination of Indigenous Peoples. It is also prescriptive, as it spells out the obligations of states to act and promote indigenous human rights, as well as refrain from practices that challenge their autonomy. Today, lawyers fight violations in international courts alongside Indigenous Peoples to enforce legal rights to Indigenous participation in decision making processes so they can autonomously govern themselves and their own ecosystems.

Case Study: Indigenous Peoples, Oil and the Case of Ecuador

Ecuador has been a hotspot of Indigenous human rights and resource conflict because the state wanted the oil that is underneath the Amazon where Indigenous Peoples are living. Oil drilling and extraction is one of the largest threats to both Indigenous communities and the Amazon ecosystem. The extraction of oil completely destroys the rainforest through chemical contamination and deforestation. Ecuador's 4.6 billion barrels of oil reserves under the Amazon are among the largest in Latin America. The state has found ways to drill under Indigenous Peoples' land by claiming that although Indigenous People are protected to live in the Amazon, the subsurface rights (everything underground) belongs to the government. Oil already accounts for a large part of Ecuador's exports and the state is economically dependent on oil drilling and extraction by foreign companies. This causes a dynamic where the state prioritizes the revenue of oil drilling ahead of the human rights of Indigenous Peoples in the state.

One of the first high profile Indigenous vs. oil company lawsuits was the $1 billion lawsuit against the US based Chevron Texaco Corporation, filed by the Indigenous People of Ecuador of Lago Agrio in 1993. From 1971 to 1992, Chevron dumped toxic wastewater – contaminated with oil, heavy metals, and carcinogens – into the Amazonian forest in open waste pits, estuaries, and rivers in a 2,000 square mile area where an Indigenous community of 30,000 was living. The pipeline ruptured 27 times by 1989, spilling 16.8 million gallons of crude.

The Indigenous Peoples teamed up with American lawyers that collectively represented more than 30,000 affected Ecuadorian Indigenous Peoples. They filed a lawsuit accusing Chevron of dumping billions of gallons of waste into open, unlined pits which contaminated soil and water and gravely affected the health of neighboring communities. The case bounced around different courts for over ten years, but in February 2011, an Ecuadorian court ordered Chevron to pay $8 billion in compensation. This was overturned in March 2014, when a United States court found that the verdict was obtained by the plaintiffs against Chevron through "coercion, bribery, money laundering and other misconduct." The case moved to an international tribunal in The Hague, which ruled in favor of the US oil company, a huge upset for the lawyers, activists, and Indigenous Peoples involved in the case.

Despite this loss against Chevron, this was not the last indigenous lawsuit against an oil company, and in some cases, and there have been great wins and success stories In 1996, the Kichwa Indigenous Peoples of Sarayaku began protesting the state of Ecuador, which, without informing the community, had signed a contract with an oil company sanctioning exploration of 65% of Sarayaku's ancestral territory. In response, international lawyers and the Kichwa people filed a lawsuit against the state of Ecuador in the Inter-American Court of Human Rights (IACHR) for violations of legal human rights. On June 27, 2012, the IACHR ruled in favor of the Sarayaku in the case of Sarayaku v. Ecuador. The court held, unanimously, that the "State of Ecuador was responsible for the violation of the rights to consultation, to indigenous communal property, and to cultural identity, in the terms of Article 21 (of the American Convention of Human Rights) to the detriment of the Kichwa Indigenous People of Sarayaku for granting a permit to a private oil company to carry out oil exploration activities in its territory from the late 1990s, without previously consulting the Sarayaku." The Kichwa argued that Ecuador let the oil company use their land to drill without Free, Prior and Informed Consent (FPIC), an established principle of many Indigenous human rights legal instruments. The court papers also stated that Indigenous Peoples who protested were subjected to beatings and kidnappings in 2003 and 2004 by Ecuadorian military contractors that were protecting state interest in oil.

Companies throughout the Amazon continue to build billion-dollar pipelines and the lawsuits will continue. As a response to resistance by Indigenous Peoples, some oil companies have tried to appease tribes by building schools and clinics, paying for teachers, and more. They have even employed experienced anthropologists to help cut deals because they are aware that the drilling in ecologically sensitive regions of the Amazon is very controversial.

International Agreements that Protect Indigenous Peoples

Indigenous Peoples are now more politically organized than ever and demand their rights granted to them by diverse international human rights instruments. Many international agreements uphold the principle that Indigenous Peoples must have consent regarding any extractive or developmental activities that may take place on their land. Indigenous Peoples rely on international human rights agreements in cases where the nation-states in which they live refuse to officially acknowledge indigenous ancestral lands. Many of these laws only take effect when they are pursued in an international lawsuit, which can be very expensive and require the mobilization of multiple resources and partner organizations.

International Agreements that Protect Indigenous Human Rights to the Environment

1) The Charter of the Organization of American States. This agreement launched The Inter-American human rights system which includes both the Commission and the Court on/of Human Rights.

2) The United Nations Covenant on Civil and Political Rights. In this agreement, the right to self-determination is protected, meaning that human beings, individually and collectively, have the right to be in control of their own destiny and to utilize and control their own land and natural resources. According to Article 27, states must take affirmative measures to prioritize – regardless of whether the state recognizes such rights of Indigenous Peoples to control their communal lands.

3) The United Nations Human Rights Committee. This agreement has interpreted a norm specific to Indigenous Peoples: "Culture manifests itself in many forms, including a particular way of life associated with land resources, especially in the case of Indigenous Peoples."

4) The International Labor Organization. In this agreement, Article 169 clarifies that Indigenous People's right to secure consultation and participation extends even to decisions about subsurface minerals or resources that remain under state ownership. This agreement also states that Indigenous communities may migrate over time and have overlapping land use. Article 6 also states:

6.1(a) government shall consult the peoples concerned, through appropriate procedures and in particular through their representative institutions

6.1(b) establish means by which these peoples can freely participate

5) American Declaration on the Rights and Duties of Man. According to this agreement, states must take these active measures: (a) Consultation must be conducted at an opportune time, (b) All necessary information must be at the disposal of all the parties beforehand, and (c) Consultation must be carried out with full consideration of the customary traditions and decision-making institution of the peoples affected.

Indigenous Peoples have formed their own political organizations to defend their international political, social, and cultural rights to preserve their environments. In the early 1980s, new indigenous organizations emerged in an attempt to develop a consensus across state boundaries. One of these groups was the Coordinator of Indigenous Organizations of the Amazon River Basin (COICA). The COICA organization is supported by UN Declarations on Indigenous Peoples. They have released statements such as the following: "Indigenous Peoples shall not be forcibly removed from their lands or territories. No relocation shall take place without the free and informed consent of the Indigenous Peoples concerned."

It is important to recognize that COICA does not see Indigenous human rights as parallel to the interests of environmentalists. A 1989 COICA letter was included in an article titled "Two Agendas on Amazon Development" in Cultural Survival Quarterly Magazine. The letter published in the article states to the global community: "We never delegated any power of representation to the environmentalist community... We want to represent ourselves and our interests directly in all negotiations concerning the future of our Amazonian homeland... We want you to recognize and protect our rights as Indigenous Peoples as we have been defining those rights within the UN Working Group for Indigenous Peoples."

SOURCES FOR AUTHOR'S NOTE

Yale School of Forestry and Environmental Studies, "The Amazon Basin Forest, Global Forest Atlas," n.d.

Jonathan Watts, "Brazilian Amazon deforestation surges to break August records," The Guardian, August 27, 2019

S. L. Lewis et al, "The 2010 Amazon Drought," Science 331, no. 6017 (2011): 554. 10.1126/science.1200807.

Jorge L. Ortiz, "The Amazon hasn't stopped burning. There were 19,925 fire outbreaks last month, and 'more fires' are in the Future," USA Today, October 18, 2019

Oliver L. Phillips et al, "Report: Drought Sensitivity of the Amazon Rainforest," Science 323, no. 5919 (2009): 1344-1347. 10.1126/science.1164033.

A. Baccini et al, "Tropical forests are a net carbon source based on aboveground measurements of gain and loss," Science 358, no. 6360 (2017): 230-234. DOI: 10.1126/science.aam5962.

Thomas E. Lovejoy and Carlos Nobre, "Amazon Tipping Point," Science Advances 5, no. 12 (2019). DOI: 10.1126/sciadv.aba2949. Lovejoy is a scientist and university professor who has worked in the Amazon for 40 years, has served as the president of the Heinz Center for Science, Economics and the Environment, and has been a Senior Fellow at the United Nations Foundation.

Richard Betts, "Fair Deal for Forests," Meteorological Office, Hadley Centre, UK.

Richard Betts et al, "Forests and Emissions: A Contribution to the Eliasch Review," prepared for the Office of Climate Change by the Met Office Hadley Centre, United Kingdom. July 2008.

Carlos Afonso Nobre and Laura de Simone Borma, "'Tipping points' for the Amazon forest," Current Opinion in Environmental Sustainability 1, no. 1 (2009): 28-36.

Douglas Sanders, "Indigenous Peoples: Issues of Definition," International Journal of Cultural Property 8, no. 1 (1999): 4-13.

Abby Ellin, "Suit Says Chevron Texaco Duped Poisons in Ecuador," The New York Times, May 8, 2003.

Juan Forero, "Texaco Goes on Trial in Ecuador Pollution Case," The New York Times, October 23, 2003.

Inter-American Court of Human Rights, Case of the Kichwa Indigenous People of Sarayaku v. Ecuador, 2012.

Cultural Survival Quarterly Magazine, "Two Agendas on Amazon Development," December 1989

ABOUT THE AUTHOR

Mona Shomali is an Iranian American author and visual artist driven by a singular purpose – to explore how we navigate competing morals and worldviews. Her unique background, steeped in Indigenous Human Rights and Environmental Science, informs her brush and her prose.

Mona's journey into the heart of the Amazon started with her pivotal research work on the Sarayacu vs. Ecuador case in the Inter-American Court of Human Rights. She worked on behalf of the Sarayacu tribe, who was suing the government of Ecuador for illegally allowing drilling on their ancestral land without consent. Having lived and worked with tribes in Guyana and Brazil, Mona has dedicated her career to teaching and sharing the crucial juxtapositions and cultural challenges inherent in Western environmentalism.

As an adjunct faculty member at The New School, Pace, and NYU, her courses focused on Indigenous human rights, resource conflicts in the Amazon, International environmental governance, and environmental justice. She also taught ecology at the New York Botanical Garden.

Mona's paintings – like those in her "Invisible Stains" collection, which includes the cover art for Water Mamas – embody her commitment to these vital global conversations.

Mona is now based in Oakland, CA. To read more of her writing or see her artwork, visit her website (monashomali.com) or her substack (monashomali.substack.com).

www.ingramcontent.com/pod-product-compliance
Lightning Source LLC
LaVergne TN
LVHW100307210926
848480LV00003B/1049
9798218838676